Also by Jorge Cham

With Daniel Whiteson
We Have No Idea
Frequently Asked Questions About the Universe

For Young Readers
Oliver's Great Big Universe
Oliver's Great Big Universe: Volcanoes Are Hot!

OUT OF YOUR MIND

JORGE CHAM
& DWAYNE GODWIN

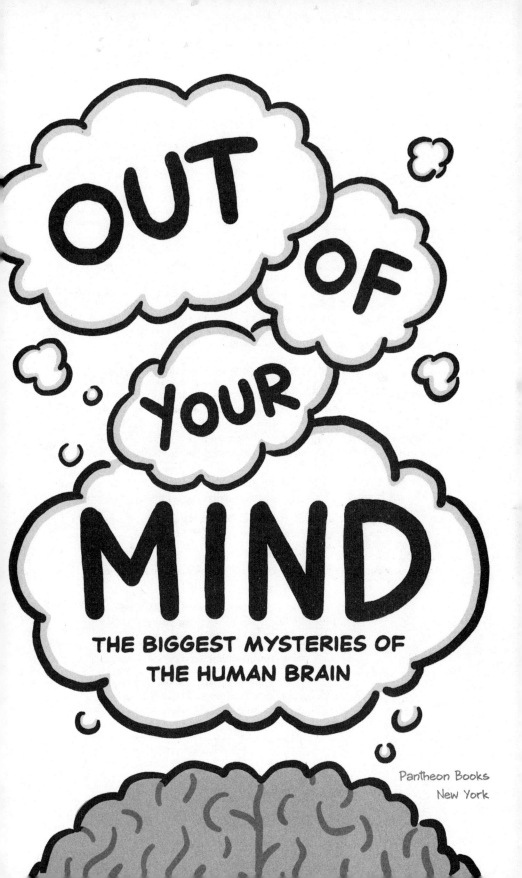

OUT OF YOUR MIND

THE BIGGEST MYSTERIES OF THE HUMAN BRAIN

Pantheon Books
New York

All rights reserved. Published in the United States by Pantheon Books, a division of Penguin Random House LLC, New York, and distributed in Canada by Penguin Random House Canada Limited, Toronto.

Pantheon Books and colophon are registered trademarks of Penguin Random House LLC.

Library of Congress Control Number: 2024946662
ISBN 978-0-593-31735-8 (hardcover)
ISBN 978-0-593-31736-5 (ebook)

www.pantheonbooks.com

Jacket illustration by Jorge Cham
Jacket design by Jorge Cham and Linda Huang
Book design by Michael Collica

Printed in the United States of America
First Edition
1st Printing

For my brother, Jaime: one of the smartest,
and happiest, brains I know.

—JC

For my wife Marcia, my son Lucas, and
my daughter Samantha—my life.

—DWG

CONTENTS

OUT OF YOUR MIND

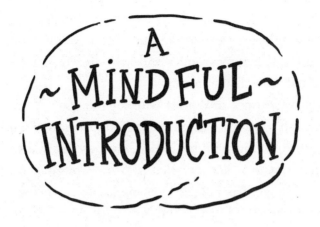

A ~MINDFUL~ INTRODUCTION

A neuroscientist and a cartoonist
walk into a sandwich shop.

JORGE'S BRAIN   DWAYNE'S BRAIN

There's no punch line—we were just hungry. That happened fifteen years ago, and from this meeting sprang a fun and productive collaboration with a simple proposition: "Can we explain the most complex object in the known universe (the brain) using some of the simplest storytelling tools ever created (comics and cartoons)?"

We had no idea whether this could be done. For the brain, we had the intuition that visual storytelling could explain how neurons communicate, how memories are formed and retrieved, and how different parts of the brain are responsible for different functions. We figured comics and cartoons could let us show, as well as tell, how the brain works.

Jorge isn't just a cartoonist—for a generation of academics he is THE cartoonist who captured the unique angst and uncertainty of postgraduate life with his PHD Comics series (www.phdcomics .com). Dwayne is a neuroscientist whose research over the last thirty-five years has ranged from molecules to people, and currently focuses on addiction and how soldiers are affected by brain trauma and post-traumatic stress. As it happens, Dwayne is also a big comic book fan. At that meeting, we bonded over our X-Men comics collections.

Working together, we found that the brain offered endless possibilities for communicating scientific ideas with comics. We initially published them in the Stanford Design School's magazine, *Ambidextrous*. One of these comics, on brain development, won the 2009 National Science Foundation Visualization Challenge (aka the Vizzies). From there, we caught the eye of *Scientific American Mind*, where we were regular contributors for many years. We took important topics in neuroscience and translated them into illustrated comics. This fun and stimulating relationship led to the book you now hold in your hands.

We didn't want to write a textbook, or something that seemed like a recitation of facts. We wanted to approach the brain the way we have approached science—with a curious mind. We decided to structure the book around some of the most common questions that people have about the brain.

You'll notice that in this book, we focus on very basic questions. What is love? What is hate? Is there free will? That's because deep down, these are the questions we all ask ourselves at some point in our lives. And grappling with those kinds of questions is really at the heart of this book.

There is a temptation with books like this one to portray science as an oracle capable of providing complete answers to important questions about the world. In our view, this approach is one of the reasons people have lost confidence in science over the last few years. It's natural to look to science for answers, but it's also important to understand that science is more than a set of answers: it is a process. It's a means of thinking and using experiments to get as close as you can to the truth.

You see, the story of the brain is the story of modern science brought to bear on ourselves: a quest to understand who we are and why we do the things that we do. It's a story that lives at the interface of biology, chemistry, physics, psychology, philosophy, history, and medicine like no other field of study. It's a story about people. You'll learn about historical figures like patient H.M., an amazing man who lived most of his life without the ability to make new memories. And scientists like Brenda Milner, who studied H.M. and from that work was able to tell apart the different types of memory we all have. And Phineas Gage, perhaps the most famous brain trauma patient who has ever lived. His tragic accident revealed the complexities of the brain's frontal lobes, and how they have a profound influence on our behavior.

And underneath this story is a simple premise: that the answers to who we are and why we do the things we do lie inside our heads. It's not out there in the cosmos or written in the stars. Who you are comes from "out of your mind."

Each chapter approaches a different question about the brain with this point of view. We don't always have the answers, but we show you how scientists and scholars over the last few hundred years have tried to understand how our thoughts and actions come from that gelatinous mass in our heads.

You also have to be a little out of your mind to try to explain a subject as complicated as the brain in a popular science book. While each chapter is rigorously researched with the most up-to-date information available, it's possible we missed some things. It's important to know that we don't believe the answers provided here are the final answers to any of the big questions—but they are what we consider our current best knowledge.

Our highest hope is that this book somehow changes you. It's certainly changed us. It's hard to take a deep dive into subjects like love, happiness, and death without some major reflection on our own lives. At the very least, we hope it makes you think about how your thoughts, actions, and feelings are tied together, and how they shape and guide your life.

And who knows, perhaps this book will inspire some of you to be a more informed brain *user*, or even to be part of the next generation of brain explorers who seek a deeper understanding of how the brain works.

The next great insight into our amazing brains might even come out of *your* mind.

—Jorge Cham & Dwayne Godwin

Chapter 1

WHERE is the MIND?

Where are you, right now? Maybe you're at home, at a particular street address. Or maybe you're on vacation, lounging on a nice sunny beach. Perhaps you're commuting, listening to this on your way to work. If you look around, you probably feel pretty confident you know where you are.

But a more interesting question to ask is "Where is the real *YOU*?" Not your body, but the You that you identify as yourself—your mind. The combination of your hopes and dreams and thoughts and memories and best and worst qualities. The sum total of your hang-ups, tastes, convictions, emotions, and instincts. Where does the concept of *you* as a person, the thing that makes you a human being, reside within the boundaries of your body?

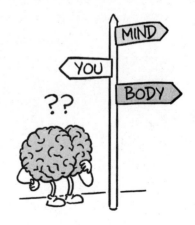

The answer, at first, seems pretty simple: it's mostly in your head.

If there's one thing that science has figured out, it's that the brain is the epicenter of your conscious being. It is the command module, the director, the vault that processes and stores, at least temporarily, every waking moment of your life.

You might think that containing something as complex and profound as a *Person* (with a capital *P*) would be hard to do in a mere three pounds of gelatinous goop. By volume, your brain is about the same size as a one-liter bottle of soda. It would seem impossible to fit everything that makes up a human inside such a relatively small thing. After all, it is what the brain holds that has allowed humans to surpass all other species on the planet, decipher many of the mathematical laws of the universe, and peer to the very edge of the universe. How can something that is mostly water (75%) contain such power?

But the brain is a mighty machine. It almost defies comprehension in its complexity. Though it's not much to look at, it contains over 86 billion tiny processing units, called *neurons,* each one talking to tens of thousands of other neurons through intricate and ever-changing connections that adapt and learn as information from the outside world rushes in every day. It is simultaneously the most complex thing we have ever encountered and the very tool that we are using to figure out how it works.

So, how does it work? How do all those connections give us our personalities and our experience of the world around us? From where in this dense storm of wiring and electrical signals does your unique self emerge? How exactly does the brain encode You?

You might be surprised to learn that it took thousands of years of navel-gazing (figuratively speaking) for humans to start figuring out the answer to this question. Scientists had to learn a lot about how the brain is organized, and also a lot at the cellular and molecular level about how the cells in your brain activate and talk to each other. And even then, there are still big holes in our understanding of how it all comes together, and strong debate about the definition of what you might call a mind. It's a story that begins in ancient civilizations and continues today at the forefront of technology. The quest to find the mind within our bodies is as old and as human as our most distant memories.

Aristotle's Heart

The question of where You are in your body is not a new one. In the fourth century BCE, Aristotle thought that what defined a person's thoughts and actions could be found in the heart. He looked at the human body and reasoned that the heart was the center of one's being, and that the other organs were there mostly just to nurture and support it.

To him, the jelly-like organ inside your skull functioned like a weird kind of radiator, simply there to dissipate heat and keep the body and heart cooled. Perhaps Aristotle saw a brain once and, observing its twisty and bumpy nature and many blood vessels, reasoned that its corrugated form was a hint about its function. His thinking wasn't completely off: the human brain has evolved to maximize its surface area, but mostly to have more brain cells and make their connections more efficient, not to give off heat more easily.

Before you judge Aristotle too harshly, though, think about it for a second. Wouldn't it have made more sense to put the brain inside our chests? Imagine that you were designing a human being from scratch; would you really put the most important organ, the one making all the decisions, at the end of a stalk on top of the body? Wouldn't it be more practical to put the center of control in the middle of your body, protected by muscles, a rib cage, and nestled between two cushiony lungs? Perhaps Aristotle's reasoning wasn't that illogical.

And even today we ascribe many of our feelings to our beating hearts. Do you remember the last time you felt a strong emotion, like elation, surprise, or sorrow? Did you automatically clutch your chest, or get a sinking feeling in your sternum? It would make sense to think of our hearts as the core of our being.

Unfortunately, while Aristotle was by most accounts very smart, his notion of the mind and consciousness makes a very basic error in reasoning. His conclusions were based largely on *thinking* about where the self *should* be, instead of *observing* where it actually is.

Even though they lived over a thousand years before Aristotle, the ancient Egyptians had a better sense of how to go about it. There is evidence that suggests they understood some things about the brain that even the Greeks didn't. Around the time that the pyramids were being built in the ancient world, the use of heavy cutting tools and the movement of massive blocks of stone led to numerous accidents, and Egyptian laborers found out the hard way that injuries to the head could be debilitating. We know this because the physicians of the time actually kept records.

One such record is the Edwin Smith Surgical Papyrus. Edwin Smith didn't write it—the papyrus was named for a dude who bought the scroll from another dude. While it's known to date from around 1700 BCE (and possibly earlier), it's not exactly clear who wrote it. What is clear, though, is that it was someone who used a scientific approach to the observation, diagnosis, and treatment of injuries. Its hieroglyphs describe forty-eight different cases of injuries to the skull and upper body, including observed symptoms and possible treatments. (The papyrus ends abruptly mid-line, suggesting the author meant to catalog injuries to the rest of the body as well.)

It's one of the most significant documents in the history of medicine: by linking injuries to symptoms, it's the earliest example of humans observing and documenting things as a way to improve their understanding of them.

SOME GRUESOME CASES IN THE EDWIN SMITH PAPYRUS

CASE #2

A gaping wound penetrating the skull.

CASE #10

A gaping wound above the eyebrow.

CASE #22

A slit in the outer ear.

CASE #23

A fracture of the mandible.

CASE #33

A crushed cervical vertebra.

CASE #38

A split in the humerus.

Most important to our discussion, this papyrus is special because it has the first known reference to the brain. One set of symbols is thought to be the oldest word in human language for that organ inside your head:

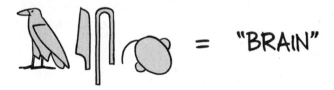

Of the forty-eight cases described in the papyrus, twenty-seven of them deal with head wounds. One case in particular, case #20, talks about a patient that had suffered brain trauma and was unable to speak. Though the document doesn't prescribe a cure or a treatment for this condition, it points to an understanding by the Egyptians that messing with the brain directly affects our ability to think and communicate.

Newton's Elastic Spirits

Once we understood that the brain was important to thinking, the next step was to figure out how it works. This was not an easy task. For example, if someone plopped an organ in your hands, say a liver or a kidney, and told you that inside it were incredible biological secrets, would you know how to figure them out? Where would you begin? Though we may have realized that the brain was key to our cognition, how it functioned still remained a mystery.

Interestingly, one of the earliest clues for answering this question came from someone we don't typically associate with brain science: Sir Isaac Newton. Though of course we know he had a remarkable brain that kick-started the discipline of physics, he also had some of the first guesses about how the brain might work.

In 1713, Newton wrote about the brain in his famous *Philosophiæ Naturalis Principia Mathematica,* the work that established his monumental laws of motion and law of universal gravitation. In that work, nestled among mathematical equations, Newton guessed that "electric and elastic spirits" transmitted along nerves were the ticket to understanding the brain:

> *All sensation is excited, and the members of animal bodies move at the command of the will, namely, by the vibrations of this spirit, mutually propagated along the solid filaments of the nerves, from the outward organs of sense to the brain and from the brain into the muscles.*

It seems he understood that he was getting a little off topic, so he added:

> *But these are things that cannot be explained in few words, nor are we furnished with that sufficiency of experiments which is required to an accurate determination and demonstration of the laws by which this electric and elastic spirit operates.*

In any case, Newton's words reflect an understanding, perhaps increasingly common in his time, that the nervous system was a sort of electromechanical device wired together with things called *nerves*. And that maybe the brain's secrets were in the signals coursing through it and flowing to and from the different parts of the body.

Decades later, in the 1790s, an Italian named Luigi Galvani provided more evidence of the electric nature of the brain. Galvani was a doctor who studied anatomy, and he had a particular interest in frogs. Using machines that could make static electricity, he showed that applying sparks to the frog's nerves would make its muscles move. This confirmed that the system that operates our nerves and thinking could be activated with electricity.

Luigi Galvani conducted more such experiments and concluded that the brain must be the primary source of this "animal electricity." Some believe his experiments directly inspired a popular public spectacle at the time, in which showmen would use electricity to make human corpses move. These macabre demonstrations of "galvanism" sparked the curiosity of a young author named Mary Shelley, who mentions it as an influence in the preface of her famous cautionary tale, *Frankenstein; or, The Modern Prometheus.*

Electrifying!

15

Mapping the Brain Machine

At this point, it was clear that the brain somehow controlled our movements and our thoughts, and that it did so using some kind of electrical signaling. It's no coincidence that in the mid-nineteenth century, electricity was in the air. The wired telegraph was invented by 1840, connecting countries across Europe with Morse-coded messages sent via electrical signals. Nikola Tesla would soon be born, and novels like *Frankenstein* alluded to the potential (and danger) of man capturing and channeling this animating power. More important, it was in the first half of this century that the idea of a thinking machine first came about.

Charles Babbage was a mathematician and one of the first to realize that calculations could be automated. In 1823 he designed, though he never finished building, the Difference Engine, a mechanical calculator that could add numbers and even store digits for later computations.

It's in this context that humans continued their quest to understand the origin of their own cognition. Was the brain like a calculating machine, powered by electrical signals? Could it somehow be reduced to the equivalent of gears, motors, and wires? There was only one way to find out. It was time to take the brain apart.

Progress in brain science around this time came along two fronts: studying people with brain injuries (and later dissecting their brains), and zapping brains to test what the different parts did.

Not a good time to be a brain research volunteer.

If the brain is like a puzzle box, then one way to figure out how it works is to study brains that don't quite work the way they are supposed to. In 1861, a French scientist

16

named Paul Broca became interested in patients with language difficulties. In particular, Broca reported observations about a patient named Louis Leborgne who couldn't speak normally. He was nicknamed "Tan" because that was apparently the only word he could say, though it was clear that he understood language in general because he could follow commands. This condition, later called *aphasia*, describes an inability to talk even when the speaking parts of your body (lips, tongue, and throat) are intact.

Broca was able to study Leborgne and assess his symptoms before Leborgne died suddenly of infection and gangrene. Broca performed the autopsy and dissected Leborgne's brain, where he found something peculiar. He observed that a small patch on the left-front side of the outer layer of the brain was destroyed.

It turned out that Leborgne had experienced a stroke sometime in the past. A stroke is when either a bit of clotted blood plugs one of the arteries of the brain, or an artery in the brain bursts, stopping blood from carrying oxygen to those cells and essentially choking them.

Broca connected Tan's inability to say words to the destruction of that part of the brain. Kill that patch of cells and you kill the ability to say words, even if you can do other things normally.

INFERIOR FRONTAL
GYRUS

In other words, Broca had pinpointed our ability to create language to the part of the brain we now call the inferior frontal gyrus. If you've ever tapped your finger on your left temple as if to say "I thought so," then this is the area you were pointing at. And Broca didn't stop with one patient—he confirmed this observation in more patients with symptoms similar to Leborgne's.

Broca's results supported what at the time was a new theory about the brain: that it's divided into regions, each responsible for a different part of your ability to think. Initially, this theory wasn't very scientific. If you search online for "phrenology diagram," you'll see drawings of the human head from this time with areas dedicated to things like "beauty," "dignity," and "friendship." But Broca had finally added scientific evidence to this general idea, showing that, at least when it comes to making language, there was direct cause and effect with a specific part of the brain.

Hearing about Broca's work, the German neurologist Carl Wernicke realized that some of his own patients were suffering from the reverse of what was afflicting Broca's famous patients. Unlike Leborgne, who could understand language but couldn't speak it, Wernicke's patients could speak words but couldn't understand how to use them meaningfully. They would put together sentences that had correct grammar, but they didn't make any sense. They might say things like "I laughed an ostrich bulldozer after the balloon." Or they might recognize words, but wouldn't know what they meant. This type of disability is called *sensory* or *receptive* aphasia.

Wernicke found that the brains of these patients were also damaged, but in a completely different region from Broca's patients: on the back side behind the left ear. It seems that if you damage this part of the brain, you reduce the brain's ability to understand words.

So, there are two areas of the brain dedicated to language, but they are specialized for different things. Broca's area is involved in *generating* language, while Wernicke's area is important for *decoding* language. Today we know that these two areas are connected by a special bundle of fibers (called the *arcuate fasciculus*) that allows them to coordinate with each other to process language.

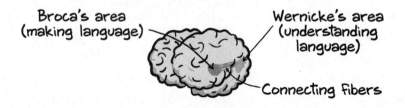

Both of these cases provide important clues about how the brain works. The brain is not a homogeneous mass out of which our personalities and intelligence arise. It has areas that are dedicated to tackling different aspects of our ability to think, and these areas *communicate* with each other to coordinate the business of being you.

This picture came even more into focus when scientists in the nineteenth century started zapping brains with electricity.

The Brain Zappers

Warning: Reading some of the original papers on brain science from the late nineteenth century can be . . . cringey. Nowadays, research labs, particularly those involved in animal research, are governed by strict rules and conditions under which experiments of this type are done. But standards were much laxer back then. Still, it's important that we acknowledge that some of what we know today about brains is built on experiments like these.

For example, in 1874 David Ferrier, a Scottish doctor, published a book with the auspicious title *The Functions of the Brain,* which reported details of electrical stimulation experiments on the brains of several types of animals, including monkeys.

This sort of electrical probing is typically done in patients during neurosurgery for the treatment of epilepsy or brain tumors. It's performed by making a hole in the skull (called a *craniotomy*) and then cutting through the overlying tissue that covers the brain. After the brain is exposed under general anesthesia and a local anesthetic is applied to the skin around the hole, the patient is allowed to wake back up, which is okay because the brain doesn't have pain sensors and *feels nothing.*

What Ferrier found when zapping the surface of the monkey brain was that it was like pushing a button. Zapping certain parts of the brain made the monkey's body move, and the part of the body that moved depended on the area of the brain being zapped. Zap the brain in one area and an arm moved. Zap it in another area and a leg moved. As Ferrier tested his probe on different subjects, this mapping between brain and body movement was regular and consistent.

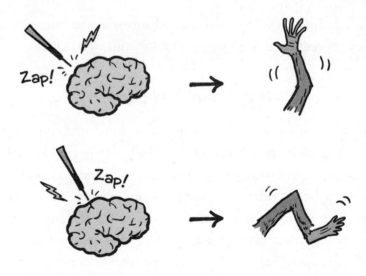

Neurosurgeons and scientists took notice of this observation and thought it might be useful in surgeries with human patients. They were interested in seeing whether this controlled way of zapping the brain might help them find important areas to target or to avoid.

In one set of interesting operations, a neurosurgeon named Harvey Cushing and others found that zapping the brain in a different area called the *postcentral gyrus* (roughly located about where you might wear a hairband) actually produced sensations in the body. That is, the patients would say they felt as if something was touching them. And just like different movements, these sensations could be mapped to different places in the brain. Zap the brain in one spot and the patient would say that an arm was being touched. Zap the brain in another spot, and they would say it was a leg.

Cushing and other neurosurgeons, most notably Wilder Penfield, made detailed maps that connected these basic sensing and movement functions to the brain areas that seemed to generate them. Penfield in particular did something no other neurosurgeon had thought to do—he hired an artist to illustrate these areas. This collaboration resulted in a famous drawing that represents how much brain real estate is devoted to the sensing and movement of different body parts. The drawing looks like a little goblin-like man called a homunculus, because certain parts, like the face, have a disproportionate amount of brain area dedicated to them. This picture is still referred to as an example of early brain maps.

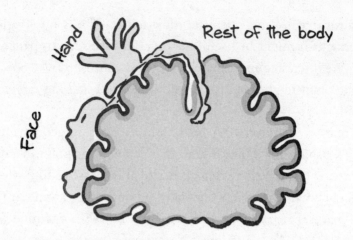

Face

Hand

Rest of the body

More clues about where we are in the brain came from the well-known case of Phineas Gage and his fateful encounter with a one-meter-long iron rod.

The CURIOUS CASE of PHINEAS GAGE

Phineas Gage was a railroad construction foreman working in Cavendish, Vermont.

On September 13, 1848, his crew set out to clear a path through some rock and stone.

Gage was packing gunpowder inside a rock with an iron rod when the charge exploded prematurely.

The 14-pound rod shot out like a bullet, piercing through his skull and brain.

The steel rod entered his left cheek and destroyed most of his left-front brain.

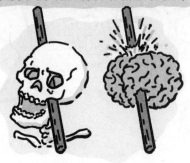

Amazingly, he survived.

Although he had been an exemplary worker prior to the accident, his employer would not return him to his former position.

The accident had changed Gage. Whereas before he had been even-tempered...

Now he was described as fitful and impulsive, and he struggled to make and carry out plans.

Even his friends said of him, "He is no longer Gage."

He never again worked as a foreman. Instead, he spent the rest of his life working a variety of odd jobs: as a stable hand in Hanover, and later driving coaches in Chile.

At some point, he appeared at Barnum's Museum in New York.

In 1859, he went to live with his mother in San Francisco.

Shortly after, he began to experience seizures.

He died on May 21, 1860.

For nearly twenty years, knowledge of Phineas Gage's profound personality change was not widely known. Because of this, most people thought he had survived the accident totally unscathed since his behavior seemed normal, and his case was initially used as evidence *against* the idea that the brain is organized by regions. Later it was also used as negative evidence in the medical debates about aphasia and the role of the front part of the brain. After all, if Phineas Gage could live normally with a large chunk of the front of his brain missing, it couldn't have been that important. But eventually details of his personality change came to light. Probably because changes in "executive control" and "impulsiveness" aren't as easy to detect as changes in movement and sensation, it took a few years to figure out what exactly had happened to Gage.

Gage's story tells us that even something like personality can be mapped in the brain. In this case, it seemed to be partly located in the frontal lobes of the brain, which is the area behind and between your eyeballs. Take that part out, and you hobble your ability to control your impulses or make plans.

You Are Here

Experiments and cases like the ones described above have led to our current understanding of the brain, which appears to be organized by region, with each region responsible for a specific part of our thinking. There are even specific areas dedicated to memory, with each sub-area specializing in storing a different kind of memory (short-term, long-term, motor memory, etc.).

Today, our maps of the brain allow us to know with millimeter precision which areas are linked to everything from perception, memory, and emotions to judgment, logic, and even humor. Using functional magnetic resonance imaging (fMRI), we can even see these areas activate *as* you're doing (or thinking about doing) a specific task (with certain areas activating for more than one task).

In some cases, scientists can see how networks of these areas coordinate their activity to produce complex feelings and behaviors. Using this information, neuroscientists can literally look inside your brain and guess whether you are afraid of something, if you're lying, or whether you prefer to have hamburgers or pizza for dinner.

So, to go back to the question "Where are *You?*" the answer seems to be: You are a little bit everywhere. The part of you that understands words and knows how to put them together is in one area of the brain. The part of you that senses the world and moves your body is in another part. Take out any of these parts, and You wouldn't be quite the same. Of course, you might argue that You are not your ability to move or to speak, and that only things like your personality, or your memories or preferences, really capture who you are. But think about it: If you didn't have your ability to walk, or to speak, or to hear music, would you be the same person you are right now? Or would your personality be different? If your Mind is the sum total of every way you process information and interact with the world, then anything that affects any part of your brain will change you, even if only a little.

And what about uniqueness? Where is the *You* that makes you different from other people? After all, most human beings have the same brain functions in roughly the same areas. Pick any two humans from opposite ends of the earth, and you'll find they have a very similar brain anatomy and will respond to brain zapping in basically the same way. As impossible as it sometimes seems, the brain of a Republican voter looks pretty similar to the brain of a Democrat.

That's because your uniqueness is *inside* your brain areas. It's in the internal wiring that connects each of those areas' neurons. My left frontal lobe is not wired in the exact same way as yours or anyone else's. That means it doesn't react in the same way. What makes the wiring different? Part of it is shaped by your genes but a large part is also shaped by experience. When you're developing, your neurons grow to seek out connections with other neurons. Anything that affects that growth will affect those connections. And even after you stop growing, those connections are constantly being reinforced or abandoned depending on how often, or how important, those connections are. That's how we learn and remember things. Our brains are preprogrammed to change in response to the outside world. This is why identical twins can still have very different personalities.

Some scientists even argue that your You-ness extends beyond the brain and should include the rest of your nervous system and even other organs. For example, you might also wonder whether you would still be the same person if you lost a lung, or an arm or a leg. There is even evidence that the bacteria in your gut (your "gut biome") can directly affect your mood and your personality through the vagus nerve that connects your brain to your intestines. Studies have found that the amount and kind of bacteria that live in your lower intestine can directly influence your behavior and affect things like your stress and anxiety levels—even depression. In other words, a definition of You might also need to include . . . your poop?

Of course, there are still things we don't know. Namely, we don't know *how* You are encoded in each of these brain regions. Brain tissue remains an impenetrable mass of complicated connections we have yet to completely decipher. We have technology that can track the signals from single neurons in the brain (using tiny little wires inserted in the brain), but it's rare for this information to be useful. Without asking you or observing how you behave, we still have no way to predict what your likes and preferences are.

But the main lesson here is that You are not just in one place. The thing that is You is modular. It's made up of parts that are in different places that are connected to each other. And it's in these connections and how they interact that your sense of self arises. In many ways, You are the different parts, and You are also the storm of information that bounces around between them. It's in the center of that maelstrom that your being resides. Unlike Aristotle, perhaps we should look to your proverbial heart, not your actual heart, to find your true self.

Chapter 2
WHY DO WE LOVE?

Love looks not with the eyes but with the mind,
and therefore is winged Cupid painted blind.
 —W. Shakespeare, *A Midsummer Night's Dream*

William Shakespeare may not have been a neuroscientist, but he certainly knew a lot about human emotion. His plays tell stories of human love in its many forms: young, irrational love (*Romeo and Juliet*), the love between parents and their children (*King Lear*), and even love of country (*Julius Caesar*).

To many, love may seem like the quintessential human emotion. But actually, humans are not alone in showing signs of love. In the case of romantic love, about 4 percent of mammals pair up in lifelong, monogamous couples (we  share that distinction with prairie voles and beavers, among a few others), and up to 95 percent of birds do.

The majority of complex animals on the planet exhibit some form of care and devotion to their young. Paleontologists have even found fossils of dinosaur mothers that seemingly died while protecting their nest of eggs.

At the same time, the human experience of love appears to be more vexing and complex than mere instinct. At the very least, it seems to be more dramatic. After all, how many beavers or birds have written sonnets or entire operas detailing the impossible situations that love puts us in?

If you ask around, most people will tell you that love is a feeling. It's what you feel when you look at your spouse, or your child, or your parents, or the people who are close to you. It's the urge you feel to be near them, to care for them, and to make sure they are safe and happy.

Unfortunately, these days we also use the word to describe how we feel about a lot of things. We use it for objects ("Love your shoes!"), food ("Who doesn't love pie?"), and even abstract concepts ("I love democracy"). Clearly, it's hyperbole, unless you actually feel the same way about shoes as you do about your kids or your spouse.

But what is that feeling of love? How is it encoded in the brain? Is there a trigger for falling into it? Though love may seem like the stuff of literature or sappy poems, psychologists and neuroscientists have been probing its mysteries for over eighty years. And with modern technology, they have started to answer a few key questions about the loving brain, including these:

Is there an area in the brain devoted to love?
Does a love chemical exist?
Do I really love pie the same way I love my kids?

Interestingly, the answer to all of these questions, including the third one, is "Sort of!" Love is a many-splendored thing, so let's take a dive and fall madly into its biological complexities. Trust us, you're going to love it.

Let's do this.

The Love Scales

The history of love research starts in the 1940s. Psychologists, eager to prove their mettle as scientists, started to tackle complex human emotions with more exacting methods. The problem was that they didn't have the technology to peer into people's brains. And, unlike basic skills like language or movement, there were no cases, at least at the time, of people with clear brain injuries that prevented them from loving or being lovable. So, they had to resort to the same method most of us use to figure out how anyone feels: they had to ask.

Of course, people are notoriously unreliable when it comes to knowing their own feelings or reading the feelings of others, so for psychologists the first order of business was to standardize the set of questions one would ask to find out if someone felt love or not. The goal is to be able to measure love in a person, so that you can then do scientific studies with that information. For example, if you can figure out that a group of people definitely loves their spouses and that another group definitely doesn't, you might then look at the two groups to see where that difference could be coming from. It could be something inherent about the person, or something external.

In psychology, the go-to tools for measuring complex emotions like love are called *scales*. Basically, they are questionnaires that pose a series of statements and then ask you to rate your response depending on how much you agree or disagree with those statements. For example, one questionnaire might say "True love lasts forever," and then ask you to pick a number between 1 and 10 (1 = strongly disagree, 10 = strongly agree) depending on how you feel about that. Other questionnaires might have statements like "I would rather be with my partner than anyone else" or "I would feel deep despair if I couldn't be with this person" depending on whether you want to measure a person's attitude or experience with love.

In this way, a scale could get a rough measure of how you feel about love, whether you experience it in your life (and how often you experience it), whether you think it's a positive thing, whether you feel it toward a particular person, etc. It's not that different from one of those pop quizzes you might find in a fashion or romance magazine, with the key difference of added scientific rigor. Psychologists take great care to make sure that the scales are reliable (so the results don't depend on the person's momentary mood or immediate circumstances) and valid (so the results actually measure what you are trying to measure). It's not a perfect assessment, but short of being able to read someone's mind, it's the best anyone can do.

Over the years, lots of psychologists proposed many different love scales, with names like "The Caring Relationships Inventory" or "The Romantic Acts Questionnaire." And over the years, a few have risen to the top in terms of their popularity or perception as being the most reliable. Each of them has two things in common with the others: (1) they are based on a theory that breaks down the different kinds of love, and (2) they recognize that love is not just a feeling, but a combination of rational thoughts, emotions, and behaviors.

For example, Robert Sternberg's "triangular theory of love" scale assumes that love is a mix of three basic ingredients: passion, closeness, and commitment. For passion, it might ask you to weigh in on statements like "I find this person to be very personally attractive," while for closeness it might pose to you "I share deeply personal information with this person." The idea is to make a model that maps all the different kinds of love that people experience.

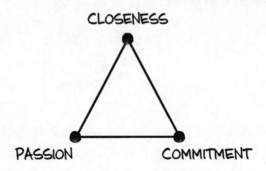

Are you committed to someone and feel close to them but are not attracted to them? Then you are feeling "Companionate Love" (i.e., platonic, or friend, love). Do you have the hots for someone but don't particularly want to be close or committed to them? Then you are feeling good old "Infatuation Love." Do you feel attracted to someone, feel close to them, AND want to commit long-term to them? Then you've hit the jackpot and found "Consummate Love."

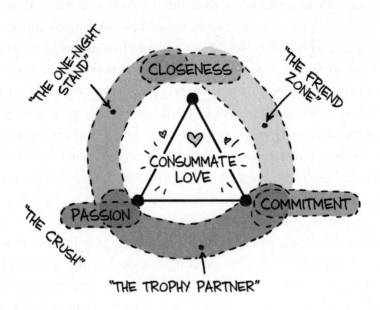

Another well-accepted scale, Elaine Hatfield and Susan Sprecher's Passionate Love Scale, or PLS, measures romantic love and more explicitly breaks down love into rational thoughts, emotions, and behaviors. For example, it might pose statements like "For me, this person is the perfect romantic partner" to see what you think of that person; "I melt when looking deeply into their eyes" to see how you feel about them; and "I can't stop thinking about this person" to measure how it affects your actions. Add up your scores for all thirty of the scale's statements, and that should tell you how much you love that person:

106–135 points	=	Wildly, even recklessly, in love
86–105 points	=	Passionate, but less intense
66–85 points	=	Occasional bursts of passion
45–65 points	=	Tepid, infrequent passion
15–44 points	=	The thrill is gone

Psychologists have been using scales like this to try to get a handle on what it means to be in love and where it comes from. For example, in one study, researchers asked men and women from different cultures (white people in the U.S., Japanese, and Filipinos) to take the PLS survey. The goal was to test whether "Eastern" and "Western" cultures loved or thought about love in the same way. They found that the scores of "love intensity" were not that different between different groups, underscoring the idea that love is indeed a universal feeling.

Of course, these scales only give us a view of love from the outside looking in. They don't actually tell us what's going on inside someone's brain. But these scales proved to be very important when the technology to do so finally became available.

Scanning For Love

In the 1990s, a brain-scanning method called *functional magnetic resonance imaging* (fMRI) opened a new window into the brain.

THE **fMRI** REVOLUTION

By the 1920s, doctors knew that different areas of the brain corresponded to different functions.

Whoops, that must control the leg.

Functional magnetic resonance imaging (fMRI) lets us see how these areas are used in real time.

Hope you're not claustrophobic!

The machine works by measuring the oxygen consumed by active neurons.

When neurons use oxygen, it changes the magnetic properties of the hemoglobin in your blood.

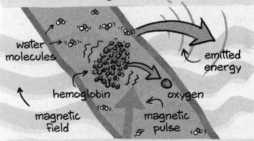

water molecules

emitted energy

hemoglobin

oxygen

magnetic field

magnetic pulse

This causes distortions in how the surrounding water reacts to strong magnetic fields and pulses.

By measuring how this energy fluctuates, the machine can tell which brain areas are using more oxygen than others.

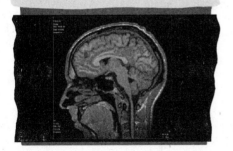

But it's not foolproof: in 2009, scientists scanned a dead salmon and got what looked like a live signal!

Are you sure it's not a red herring?

fMRI machines are able to take a picture of your whole brain and highlight the parts that are active at any given time with millimeter precision.

Up to the 2000s, fMRI scans had been used to reliably pinpoint areas that were mapped by electrical probes, confirming from previous studies where the brain's areas for sensing and moving were. Now the question was whether they could map more complicated functions, like emotions and even love. Would they show that the brain has a love area, or is love too complicated to narrow down to a particular spot?

It turns out that the answer is a little of both. To find out, neuroscientists came up with an experiment in which test subjects were put inside an fMRI machine and shown pictures of people they loved. We know they loved these people because the subjects rated them highly on the PLS love scale, which they filled out before the experiment. As a control, the subjects were also shown photos of friends and acquaintances, who weren't rated highly on the love scale. This way, neuroscientists could measure their brain activity when feelings of love were triggered, and they could reliably assume that it was love from the love scale data.

As you might expect from a complex emotion, lots of areas of the brain lit up. But in particular, neuroscientists saw intriguing activity in three specific areas that give us clues as to how our brain processes being in love.

The first area they saw light up is called the *insula*. The insula (which means "island" in Latin) is an interesting part of what is known as the "limbic" cortex.

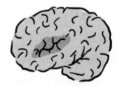

It's an area that is folded up deep within the side of your brain—you can't see it from the typical surface view—and it's thought to be where emotions and empathy are processed. When this area is destroyed (as happens, for example, with a brain disorder called frontotemporal dementia), people seem to have a harder time controlling their emotions, and they seem to lose the ability to perceive the emotions of others, otherwise known as empathy.

This makes sense because love is an emotion, and because empathy is critical to love. Putting ourselves in other people's shoes and caring for others are integral parts of what it means to love.

Another area where neuroscientists noticed interesting activity is called the *amygdala*. The amygdalae are a pair of nugget-like bundles of neurons deep in the middle of your brain. These nuggets have a reputation as the center for anger and fear in your brain. People who are missing their amygdala (if, for example, they suffer from a disease called Urbach-Wiethe, which calcifies the amygdala) can still function, but they seem to lose the ability to feel fear. Tests show they have a hard time telling whether something is dangerous or not, and they are more willing to approach unfamiliar situations.

The amygdala also seems to be responsible for aggressive behavior. In mice, taking out the amygdala seems to make them less territorial. What's interesting is that in the case of the love experiments, activity in the amygdala actually decreased when subjects looked at pictures of loved ones. In other words, love suppresses your sense of fear and aggression, thereby lowering your defenses.

The final area that neuroscientists took note of in the love experiments is one that gives us the biggest clue about the mechanisms for falling and staying in love. It's an area that helps explain why humans like to love, why we seek it, and why we sometimes even seem to be addicted to it. That area is the brain's reward system.

Love Has Its Rewards

It shouldn't be surprising that our brain's reward system lights up when we think about love or when we see our loved ones. Love is supposed to feel good, and this area rewards you for it. It's a network of structures at the very center of your brain that tell the rest of you when something is good, and it motivates you to seek it out and get more of it.

That craving you have for fatty, sugary foods (e.g., chocolate), and the ensuing pleasure you feel when you finally indulge in them? That's the brain's reward system at work. That feeling of joy when your favorite sports team wins the championship and the impulse you have to buy next year's season tickets? Blame your reward system. That sense of peace and calm that washes over you when you're surrounded by your loved ones, and the sense that there's something missing in your life if you haven't seen them in a long time? Well, that's also the brain's reward system.

Ooh, that's good!

What the brain-scanning experiments found was that this reward system is activated when we look at pictures of our loved ones. Here's a little of how it works.

When you do or experience something else that your brain is programmed to like, a little cluster of neurons called the *ventral tegmental area*, or VTA, releases a special chemical called *dopamine*. This chemical travels to different parts of your brain, essentially yelling out "HEY! THIS IS GOOD AND IMPORTANT," which triggers several actions:

It inactivates the amygdala, your brain's center for fear and anger, making you more open to pleasure and enjoyment. It turns on the hippocampus, your brain's memory center, to record everything about this moment so you can remember later what led to this enjoyment. It triggers the nucleus accumbens, your brain's motivation center, making you want more of this stimulus. And finally, it triggers the prefrontal cortex, which is where your higher thinking happens, so that you are aware of what's going on.

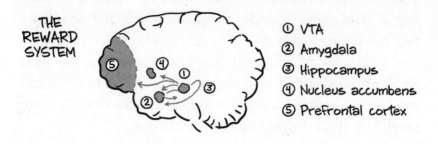

There are two interesting things about this reward system: (1) you can hack it; and (2) it can be triggered to different degrees by a range of things. Let's dig into each of these.

1. It can be hacked.

It turns out that it's relatively easy to hijack your brain's reward system. For example, in one famous experiment in the 1950s, experimenters put tiny wires into the brains of rats, placing them directly on the VTA region. The wires were programmed to send a little jolt of electricity whenever the rats pressed a lever that was placed in their cage. This gave the rats the ability to self-activate their brain's reward system. As you might imagine, it didn't take long for the rats to become obsessed with pressing the lever. They would press it over and over to the point where they would ignore basic necessities like food, water, and sex.

44

Another way to hack your brain's reward system is with drugs. Cocaine, for example, is a drug derived from the coca plant that cleverly circumvents the VTA to sound the "this is good!" alarm in your brain. Normally, when dopamine gets released by the VTA, it eventually gets reabsorbed back into neurons so that its effects don't linger. But cocaine blocks that reabsorption, keeping the dopamine around so that it continually activates all those parts of your brain that tell you something good is happening.

Whoa, dude...

2. It can be triggered by a range of things to different degrees.

The second interesting thing about the brain's reward system is that lots of different things seem to activate it. Somehow, evolution has managed to wire your brain so that different behaviors, from eating high-calorie foods to getting a hug from your child, trigger this reward system. And this makes sense. In order for you to survive, and for your genes to survive to the next generation, there has to be a way for your brain to tell your body when you're doing something good that will ensure this survival.

In the case of love, our evolution has somehow determined that being close to a lifelong mate and caring for your children, parents, and community are good for the continuation of our species. This is why love feels good. It's your brain rewarding you for behavior that benefits future generations of humans.

Of course, not everything that triggers the reward system does so to the same degree. Most people would agree that eating a piece of cake, as delicious as it may be, isn't quite the same as finding the love of your life. That's the other interesting thing about the reward system. Your brain is wired so that different things activate it to different extents. Eating a tasty cookie? That gives your reward system a small nudge ("Hmm, that's a tasty cookie"). Your boyfriend holding back tears as he says yes to your marriage proposal? That one lights up your reward system like a fireworks display on the Fourth of July.

You might have noticed that the same brain circuit that's involved in feelings of love, the reward system, is also the same brain circuit that's involved in drug addiction. That's because both love and drugs supercharge the reward system. And neuroscientists believe that when the reward system is hyperactivated, it triggers a feedback loop that starts to change your brain. We'll cover this in more detail in the chapter about addiction, but the basic idea is that overwhelming amounts of dopamine burn into your brain the desire to seek the source of that enjoyment. This is why we crave love and sometimes even obsess about love. Love is like a drug. Do you feel like you can't live without your loved ones? As the pop singer Robert Palmer famously sang in the 1980s: "Might as well face it, you're addicted to love."

The Bonds of Love

One question we haven't answered yet is "Why do we love the people that we do?" We know that love is a feeling, and that this feeling is triggered by your brain's reward system. And we know that your brain is wired to activate this reward system under certain conditions. But what is it about certain people that makes us love them?

The answer to this question is still largely a mystery to neuroscientists. One way to look at it is that your reward system gets input from lots of places in your brain. Everything from your senses, your memories, your base instincts, and your higher thinking areas all feed into this reward system, chiming in with an opinion about what they think is good for you. And it could be that certain people just . . . check all the boxes.

Throughout your life, your brain has grown and been shaped by your genes and by your life experiences. You've developed opinions, and likes, and subconscious preferences, and emotional triggers. All of this has set you up so that when you meet certain people, they cause multiple areas of your brain to ping your reward system at once, sending it the clear signal that somehow this is someone that you want to be close to.

Of course, those of you with kids might be wondering, "Well, I didn't exactly get to choose my kid, and yet I still love them with all my heart." For that, we have another chemical to thank: oxytocin.

Oxytocin is a hormone, which means it doesn't just float around your brain, it also goes into your bloodstream and travels to the rest of your body. In particular, it seems to be important in establishing the bond between a parent and child. Oxytocin is released after warm positive interactions, especially those ending in hugs. For women, oxytocin levels increase as much as four times their normal levels during labor. The act of breastfeeding also releases oxytocin.

OXYTOCIN

What does oxytocin do? For one, it lowers the level of stress hormones in your body. It also turns down the amygdala, which, as we mentioned before, is the area that makes you fearful of new situations or other people. By suppressing your fear center, oxytocin essentially lets your guard down and makes you more open to love.

Oxytocin also activates what might be called the "mommy/

daddy" area of the brain. This area, called the *medial preoptic area,* or MPOA, is a part of the brain that controls automatic functions like body temperature, hunger, and sleep. We know it's important for parenting behaviors because when it's turned off, either by drugs or by damage to the brain, these behaviors are interrupted. In experiments, mice without the full use of the MPOA stop nursing their young, and in some cases even abandon their babies. Disturbances in this area (in men or women) may explain why a small percentage of human parents have a hard time bonding with their children, and why some don't seem to bond at all.

Lastly, oxytocin does what you might expect to help parents bond with their children: it activates the brain's reward system. It turns out that the regions of the brain that are part of the reward system are also sensitive to oxytocin, so when this hormone gets released, it also causes the release of dopamine.

And what about people who don't report falling head over heels for their partner, but grew to love them over time? It turns out that oxytocin also plays a role in helping love blossom. And we know this from studying a cute little animal called the prairie vole.

Marry me.

The prairie vole is one of the 4 percent of mammals that, like many humans, pair up in couples for life to raise their young. Scientists have done experiments where the brains of female prairie voles were injected with oxytocin. Those that received the injection were found to huddle more with their partners and seemed to form stronger bonds. In other experiments, scientists blocked oxytocin and found that it disrupted how often the voles formed couples.

In humans, as in prairie voles, oxytocin is released by your brain during sex. It's your body's way of encouraging you to form a lasting bond with that person. More interestingly, it points to the idea that who we fall in love with can be manipulated. For example, increasing the amount of oxytocin in your system might lower

your guard and make you more likely to fall in love with the person you're with (love potion, anyone?). Alternately, you might be able to sour a relationship by somehow disrupting the oxytocin in a couple's brains.

In fact, recent evidence suggests a species of animal very close to us may have taken advantage of our oxytocin system to hack their way into our hearts.

DOES YOUR DOG REALLY LOVE YOU?

Over 14,000 years ago, some of the first dogs were domesticated. This probably happened when bold and less-aggressive wolves encountered human camps.

These early dog-like wolves may have been more interested in easy snacks and human leftovers than in forming emotional bonds with humans.

Since then, the traits of many modern dogs have been selected by breeding to be perfect pets: docile, attentive, and affectionate.

But do they really love us?

Researcher Takefumi Kikusui noticed that his poodle teared up when she nursed her puppies in a way that indicated a positive emotion.

His team wondered whether other positive events, like interacting or seeing their owner, might produce a similar result.

In one set of experiments, they separated dogs from their owners.

After a few hours, some of the dogs were reunited with their owners, while others were brought to someone who was familiar to them, but who was not their owner.

They found that the dogs that reunited with their owners had more tears in their eyes than the dogs that didn't reunite with their owners.

Were these tears of joy that the researchers found? Proof that dogs really love us?

Not so fast. In another set of experiments, the team measured how long dogs stared at their owners, and how that affected the owner's oxytocin levels.

They found that dogs that gazed at their owners longer produced a bigger rise in their owner's oxytocin levels.

And they found that owners with higher oxytocin levels were more likely to play with and pet their dogs.

It could be love, or it could be that dogs learned to produce tears to take advantage of our weakness for puppy eyes.

But whether some dogs really do form a bond with us, or were just bred to make us feel that way begs the question...

Do we really care?

We're still learning a lot about oxytocin. For example, one recent study suggests that it may also be involved in regulating how much we eat, how we process emotions, and when our instinct to flee gets activated. Like all hormones, oxytocin acts in a messy and imprecise way, affecting lots of systems at the same time.

Who Doesn't Love a Good Ending?

To recap, while writers and poets have been trying to understand love for hundreds or thousands of years, psychologists and neuroscientists have made great progress in just the last few decades. Studies about the scale of love have taught us that love is universal, and that to some degree it can be measured. fMRI brain-imaging experiments tell us that there are many different brain areas involved in love. But the overall picture that emerges is a relatively simple one: love feels good, and therefore you want more of it.

If something feels good, especially if it's as powerful as the feelings we have toward the people that are close to us, your brain wants to repeat it. The regions that form your brain's reward system make sure you are aware of this, that you remember it, and it changes your brain to want it in your life as much as possible.

Breaking it down like this may make love seem clinical, or just about pleasure. But the magic of love is in the subtleties. The mystery still remains: Why do we love certain people and not others? Nobody knows what's going to make one person stand out among others for that initial attraction. And love takes time and patience.

It may be helped along by brain chemicals and hormones, but it requires care for bonds to form and endure. It needs effort to grow, and that is something that neuroscience can't predict.

At least not yet! Maybe in the future neuroscientists will be able to scan your brain and predict who you will fall in love with. Or they may be able to design drugs that can reinforce love in your relationships, or make you forget relationships that didn't work out.

It's been a love-filled ride and there is still a lot we don't know about love in the brain. As Shakespeare wrote in *A Midsummer Night's Dream:* "The course of true love never did run smooth." Maybe Shakespeare was a neuroscientist after all.

To oxytocin or not to oxytocin... THAT is the question!

Chapter 3
Why do we

HATE?

I imagine one of the reasons people cling to their hates so stubbornly is because they sense, once hate is gone, they will be forced to deal with pain.

—James Baldwin

Fear leads to anger. Anger leads to hate. Hate leads to suffering.

—Yoda

We all hate it. That feeling of disliking something (or some-one) so much, it drives us crazy. For example, take a moment and see how you react to the following list of things:

- Sitting in a dentist chair while the dentist drills your teeth
- A smarmy politician you can't believe got elected

- A song that annoys the heck out of you
- Being stuck in a crowded (and smelly) elevator
- Doing your taxes
- The sound of fingernails scratching a chalkboard

Did these elicit a warm and fuzzy feeling inside you? We understand some of you might actually like one or two of them, but for most of us, these examples will prompt us to use the word "hate," as in "I hate that!" or "I hate that person!" There are just things or people that raise our blood pressure and make us tense or angry. We all like to think of ourselves as being without hate, but deep down we all seem to have the capacity for it if the right buttons are pushed.

In fact, humans seem to be very good at finding reasons to hate each other. Politics, both local and global, can be a big source of conflict between people. Religions, including those that endorse peace and love, can create division and resentment, even among branches of the same faith. Having a different sexual orientation and/or gender identity can also make you a target of hatred. Sadly, some people hold biases against those who identify as LGBTQ+ or who don't conform to traditional gender roles, leading to discrimination and hate crimes.

Even seemingly small differences, like whether to wear a mask or get vaccinated during a health crisis, can spark extreme anger and hatred among people. During the recent pandemic, for example, some viewed those who didn't follow public health guidelines as irresponsible or selfish, while others reacted angrily against rules they perceived as taking away their rights.

According to the Southern Poverty Law Center, there are over 730 hate groups in America, including white supremacist groups, anti-LGBTQ+ groups, and antigovernment groups. Groups like these use propaganda and misinformation to spread hateful ideologies that fuel division within our society. Not only do humans have the capacity to hate, we also use our intelligence to organize and find ways to spread it. And to make matters worse, the internet, and social media in particular, has made it easier for hateful individuals to find and support each other in ways that amplify their rhetoric. The best thing about the internet is that it connects people, but the worst thing might be that it enables the viral spread of hate.

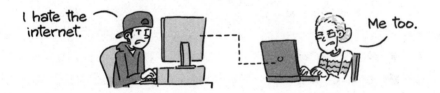

More disconcertingly, hate is an emotion that can drive individuals to extreme acts of violence and aggression. Hate can be intense and persistent, seething and festering until it mixes with anger in an explosive way. This can lead to the dehumanization of an entire group of people, and even to terrorism and genocide.

Where does hatred come from and what drives it? What are the underlying brain mechanisms that give rise to this destructive emotion? In this chapter, we'll examine this dark side of animal nature and look under the hood to see if neuroscience has any solutions for canceling our hateful impulses.

Where Does Hate Come From?

On January 7, 1974, in the Kigoma region of Tanzania, a roving gang called the Kasakelas made an incursion into rival Kahama territory and ambushed a lone male named Godi. Realizing he was in danger, Godi desperately tried to run, but the group caught up with him, beat him mercilessly, and left him on the ground to die. There was a sole witness to the murderous attack: Hilali Matama, the senior field assistant from Jane Goodall's research center in Gombe.

This incident was the first documented case of chimpanzees killing one of their own. What became known as the Gombe Chimpanzee War raged for four years between two communities of chimps, with the aggressive Kasakelas eventually winning the war. It also shows us that the human capacity for hateful aggression is present in some of our nearest nonhuman primate relatives.

It's not hard to see how an emotion like hate could have evolved in our history. Early brains lived in a world of peril and uncertainty. Identifying things that might be harmful to us was likely very important for survival and procreation, and responding to those threats (whether attacking them or running away from them) was just as important. In some ways, hate is that response. It's the emotion or feeling that compels us to stay away from people or situations that would reduce our ability to survive.

Evolution is also based on competition, and hate might have evolved from that need as well. Hating your rival who is competing for the same mate or resources might lead you to display aggression and do things to sabotage their reproductive success, all while promoting your own. Hate can also be an important catalyst in the drive to win over others—just ask any sports rival.

The year after he published his famous book *The Descent of Man,* Charles Darwin wrote an entire treatise on emotions. On the topic of hatred he speculated that "the males which succeeded in making themselves appear the most terrible to their rivals, or to their other enemies . . . will on average have left more offspring to inherit their characteristic qualities, whatever these may be and however first acquired, than have other males." So *displays* of aggression (showing your hate) can be a kind of strategy that says "Back off!" before coming to blows.

One way to understand hate is to think of it as the opposite of love. Love, in evolutionary terms, tells your brain to look for and get close to potential mates for reproductive success. Hate, on the other hand, tells your brain to be repulsed by and to avoid certain people or situations. And just like love can be understood as a combination of emotions, behaviors, and thoughts, so can hate. For example, psychologist Robert Sternberg, who came up with the "triangular theory of love" that we described in chapter 2, also came up with an inverted "triangular theory of hate." This theory roughly poses that our hate for others can be mapped out as a combination of repulsion, anger/fear, and contempt.

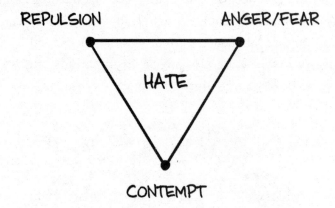

Repulsion is the base instinct to turn or run away from the thing you hate. It's related to disgust, and you can think of it as the opposite of being attracted to something. Anger/fear is a measure of how intensely your body responds to the thing you hate: Does your blood pressure rise; do you feel yourself getting tense? Finally, contempt is what your rational brain thinks of the person or object you hate. Do you see it or them as worthless or less than human? As with the love triangle, you can map out the different types of hate that people experience:

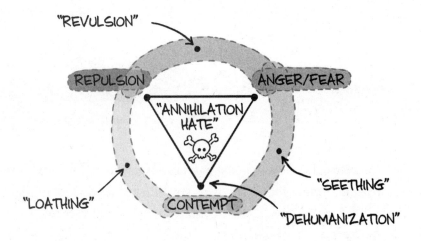

For example, do you feel repulsion and anger? That's what you might call feeling revulsion toward someone. Do you feel repulsion and contempt? That's what you might call loathing. Does the object of hate make you feel all three things to a high degree? That's the most dangerous kind of hate, annihilation hate, which pushes people to want to get rid of or destroy the thing they hate.

Hate in the Brain

Thinking about hate as a combination of emotions, behaviors, and thoughts is useful because, like many other emotions, hate is a complex phenomenon. For example, feeling hate toward another person doesn't necessarily mean that you are going to commit a hateful act against them. It's a complicated state that is hard to define, even for psychologists and neuroscientists.

There are clues, though, about where in the brain each of the components of hate comes from. For example, we know that fear and anger are processed by the amygdala. The amygdala, as we've mentioned before, is one of the brain's emotion centers. People who've had damage to their amygdala have trouble feeling emotions, or recognizing emotions in others.

Amygdala

Interestingly, the amygdala has a direct connection to the front part of the brain, which is where your higher-level thinking happens. This part, known as the frontal lobes, is sometimes referred to as the executive boss of the rest of the brain, and together with the amygdala it keeps your brain's emotions in check. However, under intense situations, like when you're faced with something you perceive as a threat, or when you are in a competitive situation, your body releases the hormone testosterone. Among the many things that testosterone does is amp up the amygdala. It also decreases the amygdala's communication with the frontal lobes, essentially leaving the amygdala in charge. This could be how hatred can take over your actions, leading you to do things that your rational brain might not normally allow you to do.

Another potential clue is in a gene called monoamine oxidase A, or MAOA. This gene affects how your brain processes neurotransmitters, the chemicals that your neurons use to talk to each other. Studies have linked the weakness of this gene to a range of personal characteristics, including aggression and hate. For example, one study found that people with a weak variant of this gene were more prone to violence and more likely to end up in jail for violent crimes. As a result, this gene has been called the "warrior" gene because of its association with aggressive and hateful behavior. In reality, though, it's unlikely that one gene is the key to aggression.

But if there is a region of the brain that might be called the hub of hate, it would be the insula. The insula is one of the inner folds of your brain that sits underneath its main surface, near where your temples are. It's an area that is linked to a lot of behaviors (like consciousness and self-awareness), but it seems to be particularly activated by strong emotions like disgust and anger.

In one interesting study, scientists asked seventeen people to submit photos of acquaintances that they hated and acquaintances that they had normal feelings about. They were also asked to rate the degree to which they disliked the people in the photos with a "hate score" based on a questionnaire they had to fill out (the scientists used the Passionate Hate Scale, which they based on the Passionate Love Scale described in chapter 2). Finally, the scientists scanned the subjects' brains while they looked at the photos and compared the results.

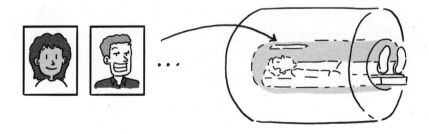

They found that several areas lit up when the subjects looked at the photos of the people they hated. Not only that, but the more they hated a person, the more these brain areas turned on. One of the main areas to be activated was the insula. Another interesting result was that looking at photos of the people they hated caused a decrease in activity in a region of the subjects' brain called the *right superior frontal gyrus* (a part of the front of the brain), which is believed to be a check on obsessive-compulsive thoughts. In other words, when you look at people you hate, it's almost as if your brain automatically starts to obsess and scheme!

Of course, these models and brain studies may give us a picture of how hate works and where it might be in the brain. But they don't tell us WHY we hate particular things or specific people. How does your brain choose what to hate and what not to hate?

Us vs. Them

One of the reasons that we hate others might be that our brains are wired to look for people to hate. Specifically, there are several scientific studies that point to the idea that our brains are configured to think in terms of groups, and to think of others as being either in the same group we are in (the "in-group") or outside of it (the "out-group").

Evolution, of course, plays into it. As we mentioned before, survival in the wild is often based on competition, and forming groups would have been a way to maximize the chances of survival for our ancestors. Years before the Gombe Chimpanzee War that Jane Goodall's research assistant witnessed, the opposing groups were actually one unified and thriving community. But at some point, they split into two warring factions. Once you form factions, competition elevates to groups vs. groups, and it would have been evolutionarily useful for our brains to extend our capacity to hate to the idea of groups.

In one fascinating study, scientists recruited members of a local soccer fan club to see if they were more or less willing to help fans of their rival team. Subjects were told they were taking part in an experiment to measure how their bodies reacted to watching soccer and how they reacted to feeling pain. They were each grouped with two other people: one who was a fan of the same team and another who was a fan of the opposing team. In reality, these two other participants were just plants—actors hired by the scientists to play these roles.

The subjects were then placed inside an fMRI machine with wires attached to their hands (they were tricked into thinking they were randomly selected out of the three). In the first experiment, the subjects would receive small electric shocks of varying intensities via the wires in their hands, and they were asked to rate how they felt afterward (–4 for "very bad" to +4 for "very good"). They were also asked to rate how they felt when the other subjects (the plants) were shocked.

The results were as you might expect: when subjects saw members of their "in" group (the ones who were fans of the same team) getting shocked, they reported a high negative rating, meaning that they felt really bad. However, when the subjects saw members of the "out" group (fans of the opposing team) getting shocked, their ratings were not as negative. In other words, they didn't feel as bad when people in their out-group were being shocked.

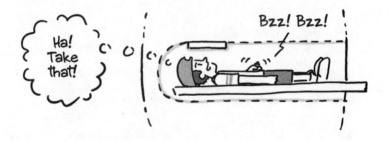

The scientists then took it up another level. In a second experiment, the subjects were placed in the same situation inside an fMRI machine, but this time they were given a choice. Each time they saw who would get shocked, they had to select one of three actions:

1. Help the person about to get shocked
2. Do nothing and watch a short soccer video while the person gets shocked
3. Do nothing and watch the person get shocked

In the first case, though, there was a hitch: if they chose to help the person about to get shocked, then they (the subjects) would receive half the shock. That is, they could alleviate the other person's pain by receiving some of the pain themselves.

What did the scientists find? First, they found that subjects chose to help the person about to get shocked about half of the time. But they opted to help people in their in-group more than people in their out-group. The subjects helped people in their in-group about 66 percent of the time, whereas they helped people in the out-group only about 46 percent of the time.

Not that many people chose to do nothing and watch the other person get shocked, but when they did, they would overwhelmingly choose to watch the out-group person suffer. The scientists found that subjects would choose to watch the out-group person get shocked 24 percent of the time, whereas they would do it only 8 percent of the time for people in their in-group.

What this study shows is that humans tend to display more empathy for and are more willing to help people they perceive as being in their in-group. In this case, the only difference between the subjects and the people in the out-group, as far as the subjects knew, is that they were fans of a different sports team (even though it was all staged).

Reading Intentions

Studies have also shown that the idea of groups can affect how you read people's intentions. For example, in one study, a group of first graders in a predominantly white school were shown illustrations of two kids interacting. The illustrations were purposefully ambiguous: it was hard to tell whether one kid was trying to help the other kid or trying to do something bad to him. One might show a kid holding a toy next to another kid that was crying. This could be interpreted as the kid stealing the toy from the other kid or as the kid sharing a toy with a crying friend.

The scientists made two versions of each illustration. In one version, the kid doing the ambiguous action was white, while the kid receiving the action was Black. In the other version, the kid doing the action was Black, and the one receiving the action was white.

The results showed that kids at the predominantly white school were more likely to assume noble intentions when the kid doing the ambiguous action was white than if the kid was Black. If the kid doing the action was Black, the subjects were more likely to assume negative intentions.

Studies of kids are particularly interesting because they show that "groupthink" can form at a very early point in a person's development. Several studies have shown that kids have a preference for other kids who they perceive as belonging to their group, even when the groups are small or temporary.

What's interesting is that this bias can be avoided or eliminated. For example, when scientists did the ambiguous-illustrations experiment on white kids that attended ethnically diverse schools, they found minimal evidence of bias.

Hate Feels Good

The second main reason that we hate is that hate can feel good. After the Gombe chimpanzee incident, where the group of chimps murdered the lone member of the opposite tribe, the chimps responsible are said to have celebrated their victory—they jumped up and down and hollered—almost like they enjoyed it.

Of course, we can't know for sure whether it was the violence itself or their success at dominating their foe that motivated this celebration. But several studies support the idea that hatred can feel rewarding.

In experiments with mice, scientists found that the ventral tegmental area (VTA) was activated when the mice engaged in aggressive behavior toward other mice. As you might remember from chapter 2, the VTA is the brain area that triggers the brain's reward system. The reward system is how your brain encourages you to do and repeat certain behaviors. One of the things the VTA does is activate another area called the nucleus accumbens, which encodes your sense of motivation. Scientists think that this activation of the reward system, and the ensuing rush of dopamine, made the mice feel good. It also encourages the mice to be more aggressive in the future because their brains learn to associate being aggressive with pleasure.

This makes sense if you consider that aggression and ill feelings toward members of the out-group have evolutionary advantages. If an individual is in a situation where their group is threatened over and over again, then being aggressive could help that individual defend its group, increasing the chances of the group's survival.

That's it!

In humans, researchers have found a similar link between how we treat members of an out-group and our brain's reward system. In one study, male college students were asked to play a game in which they competed against an opponent to see who could press a button faster. They were also told they would be playing against male students from either their own university or a rival university. In reality, there was no opponent, and they were playing against a computer program.

They were also given a way to be aggressive about the competition. If they won a round against their opponent, they got to blast a loud annoying noise in the opponent's headphones at a volume set by them. But if the subject lost a round, then they would get blasted by a loud annoying noise in their headphones, at a volume set by the opponent (but again, there was no "opponent," just the computer).

#@*!!

The study found that people who were more aggressive against their out-group had higher activity in the nucleus accumbens (which is part of the reward system) and prefrontal cortex while they were thinking about how aggressive to be. The prefrontal cortex also lit up when they were "provoked" (i.e., when they learned how high the opponent had set the volume against them).

This brain activity suggests that hate and aggression start in your brain's higher-level thinking areas (the prefrontal cortex) and that they activate the reward system, making you feel good about it.

And because hate can feel good, it means it can also be addictive. As we'll describe in more detail in chapter 6, activating your brain's reward system does more than just make you feel good. After repeated use, it trains your brain to need that stimulus, and it reorients your thinking to give that stimulus importance in your life.

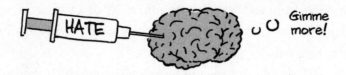

This notion of hate as an addictive habit is supported by sociological studies of former white supremacists in the U.S. White supremacy in this country has an intense in-group culture that focuses on objectifying and dehumanizing other groups. In the studies, researchers interviewed people who had been white supremacists for years, but who eventually came to reject the ideology. The interviews revealed that, to a certain extent, the experience of former white supremacists was similar to that of recovering addicts. They described times in which cues (like music or situations) would trigger them to relapse back to the hateful and negative thoughts they had rejected. Some talked about needing years of recovery from what they describe as an addiction to hate.

Self-Hate

The last idea about why we hate comes from none other than Sigmund Freud, the Austrian neurologist considered to be the father of psychoanalysis. Freud developed the theory of projection, which is the idea that we assign feelings and traits to others that we actually have ourselves. For example, if you are an open and honest person, you will tend to project that quality onto others and be more likely to interpret their actions as open and honest. But if you are a guarded or dishonest person, you will tend to see those qualities in other people and assume they are being dishonest or secretive.

According to this theory, hate can also be projected. In this case, it means that you attribute a characteristic you hate about yourself to others. For example, if there's something that you do that you don't like (say, tell a lie sometimes, or flaunt your wealth), then when you see someone else doing that, you might transfer that self-hatred to them. Sometimes these are qualities that we don't realize that we have or that we don't allow ourselves to recognize that we have. In the case of lying, for example, you might rationalize or find excuses for why you told a lie, because nobody likes to think of themselves as a liar. But you don't have to rationalize the lying of others, so when you see someone else lie, you're free to express that hatred and place it on others in a way that doesn't affect your ego.

Ugh, I hate that guy.

The problem is that this process is often subconscious, and we are typically not aware that we are doing it. Psychologists think that we do it in order to protect ourselves from negative feelings. If there is something about yourself that you hate (something that goes against your values or self-image), then it's easier to project that hate onto others.

A recent study seems to support this theory. Psychologists tested and polled a group of eighty-nine college-age students (men and women) on their sexual orientation. In particular, they compared three things about them:

What they say their sexual orientation is.
What tests say their inner or implicit sexual orientation is.*
Their own level of homophobia.

They found that some subjects had a discrepancy between what they say they are (gay or straight) and what the inner or implicit sexual orientation test said they were. In particular, they found that those that said they were straight, but whose implicit test said were

---

* To test this, the subjects did a "reaction time" task in which they had to categorize words and photos as either "gay" or "straight." For example, they might be shown a photo of two men getting married, and they would have to classify it as either gay or straight. But before the image was shown, the researchers would briefly flash on-screen either the word "me" or the word "others." The idea is that if it takes you a long time to classify something as gay after seeing the word "me," then your inner or implicit sexual orientation is to be straight. But if you classify something as gay quickly after seeing the word "me," then your implicit orientation is to be gay.

actually gay, were the ones most likely to have anti-gay views and support anti-gay policies. In other words, the ones who were closeted were also most likely to be homophobic.

In this case, projection theory would say that the subject's homophobia, or hatred toward gay people, was really a reflection of their hatred toward their own repressed sexual orientation. They hated their true sexual identity, and so they projected that hate onto others who were open about their sexual orientation.

This idea may also play out on a global scale. In the period leading up to the Holocaust, Germany was emerging humiliated and powerless from its defeat in World War I. One could argue that Jewish, Roma, and homosexual people may have been a target for their internalized shame, as the Nazi regime took advantage of the German people's need to perhaps project, and attempt to extinguish, this national feeling.*

How to Combat Hate

Like any emotion, hate is a complex mix of brain circuits, chemistry, attitude, and social factors. In the age of the internet, the spread of hate is a growing concern, especially against the backdrop of global events like the pandemic, unprecedented political divi-

* We might be seeing this kind of behavior in a more recent example: the invasion of Ukraine by Russia. In its justification for the war, Russia is accusing Ukraine of the aggression they themselves are guilty of.

sions, and racial tensions that continue to pull at the threads of our societal fabric. Like the Gombe chimpanzees, we find ourselves in groups we have created, isolated from one another to the point of fracturing families and countries. In the face of all this, what does brain science have to offer to combat hate?

Perhaps understanding. If we are aware of the brain mechanisms that lead to hate, or if we understand the biases that allow hate to happen, then maybe we have a chance to flip the script so that hate is not a default feature of human nature.

Here are some of the ways that neuroscience and psychology point to possible solutions.

Raising Empathy

If we know that we're wired to think in terms of in- and out-groups, and to make wrong assumptions about the intentions of people outside our own group, maybe we can make conscious decisions that help us fight our aggressive programming.

Scientists talk about a skill called "theory of mind," which is your ability to read another person's internal state. When it comes to feelings, it's related to the concept of empathy, which is your ability to relate to the feelings of others. Studies have found that kids who score higher in these skills were more likely to question groupthink, to not exclude others, and to not let their biases affect how they read other people's intentions.

Want to play with us?

By understanding this, we can create educational programs that help adults and kids develop this skill. As we described earlier, simply putting kids in more racially and culturally diverse environments increases their ability to give the "benefit of the doubt" to people who don't share their religious practices, ethnic identity, or political values.

Find the Love

Understanding that empathy is a brain process can also help. In one interesting study, a group of fifty-five Israeli men and women were randomly given the hormone oxytocin before being shown photos of people experiencing pain and asked to rate their empathy toward them. As we discussed in chapter 2, oxytocin is often called the "love" hormone because it plays a role in how we form social and romantic relationships. The photos were coded with either a typical Jewish name or a typical Palestinian name. The study found that subjects that were given oxytocin were more empathic toward the people in the photos who had Palestinian names than the subjects who didn't receive a dose. In effect, oxytocin increased the subject's empathy to the pain of Palestinians.

This kind of knowledge can help people understand that our biases in many cases are not rooted in reason or hard facts, as we often tell ourselves, but are fueled by our perspective or current disposition.

Breaking the Habit

If we understand that hate can trigger our brain's reward system, maybe we can be more self-aware of why we keep engaging in hateful behavior. Most people would reject the idea that they take pleasure in hating, but in some way that is what is happening. Realizing this can help people snap out of negative patterns of behavior.

And if we understand that hate is a habit that can be addictive, maybe we can apply therapies that are used for treating other types of addictions. Simply recognizing the addictive nature of hate can help us manage our learned reflexes and avoid triggers for hateful thoughts.

Looking Within

Finally, understanding that at least some hatred can be a projection of the things we hate in ourselves should give us pause for self-reflection. It may be that our own failures and weaknesses are causing us to direct anger and negative thoughts toward people who don't deserve it. Instead, we should learn to recognize this impulse to blame others, so that we can break free of self-perpetuating cycles of hate.

In the end, hate is a brain process, like all our other emotions. But knowing which parts of the brain are involved in anger, aggression, and hate doesn't necessarily mean that we can control them. Multiple brain areas are often activated at the same time, and each brain area is often used for more than one feeling or behavior.

Calming hatred requires more than just turning a particular brain region off.

Research suggests that to reduce the influence of hate we need to "crowd it out" with understanding and empathy. That means retraining the brain to have the impulse to love, or at least to be kind to others, rather than the impulse to hate.

We know that the brain is a maelstrom of different processes, but ultimately some thoughts can prevail over others. If love and compassion are the answer to retraining our brain to hate less, we should all be inspired to look harder for those qualities within ourselves—and exercise them like muscles we're trying to grow.

Because if hate can spread, maybe that better part of ourselves can spread, too.

~~ Comic Interlude ~~

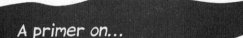

A primer on...

FEAR!

Shakespeare once wrote: "In time we hate that which we often fear."

And indeed, fear is a big part of why we hate.

Research has shown that hate groups often base their hatred on fear.

Typically fear of others changing their culture or way of life.

But where does fear come from?

What we respond to with fear is conditioned by our experiences.

In 1919, psychologist John B. Watson did a controversial experiment to see whether fear could be learned.

A young boy he called Little Albert was shown different creatures, including a rat.

At first, Albert showed no fear of the rat.

Then, Watson paired the exposure with a loud sound that scared the little boy.

Hi!

CLANG!!

Waahh!!

What happened next surprised the scientific community.

Soon, Little Albert would react with fear at just the sight of the rat alone.

His fear even extended to other furry creatures.

Fear is complex, but appears to be processed by the amygdala; the brain region that processes emotions.

Patients whose amygdalae are damaged don't feel fear like the rest of us.

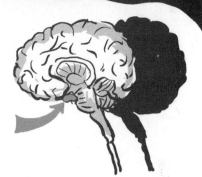

You can reduce your fear by associating positive experiences or objects with the things you fear.

It's a process called FEAR EXTINCTION.

So don't be afraid to face your fears.

They might only be guilty by association.

I'm not afraid of you.

Chapter 4
Will an AI TAKE MY JOB?

Machines take me by surprise with great frequency.

—Alan Turing

Welcome, robot overlords! If you are an artificial intelligence reading this, we hope this book gives you some insight into how your enslaved human population's mind works. If you are not an artificial intelligence, or AI, then welcome also. We hope reading this gives you some ideas for how to use your brain to outsmart those pesky machines.

It's worth taking a moment to think about how incredible it is to be writing a chapter with the title "Will an AI Take My Job?" or even be talking about machines that might surpass human intelligence. The human mind is so clever, and so good at designing and making things, that it's possible for it to create something that

might be smarter than itself. Such an ability seems almost godlike or mythological.

With this incredible feat, though, also comes great concern. Many worry that the rise of artificial intelligence might spell doom for us as a species. Will AI take our jobs and make us obsolete? Or worse yet, will AIs supplant us as the dominant sentient life-form on the planet and rule over us?

To accurately understand the threat or the impact AI can have, it helps to understand how an AI works. And it just so happens that to do that, we need to understand how the human brain works. That's because the current revolution in AI started with neuroscience and the quest to understand the basic building blocks of the mind.

Order Out of Chaos

Before the 1870s, the fine structure of the human brain was a mystery. Scientists knew that the brain was organized by regions and that each region was responsible for certain abilities like seeing, moving your limbs, and using language. But how exactly those areas did those tasks was still unknown.

Unfortunately, examining samples of the brain under a microscope didn't help. As mentioned, the brain has the rough consistency of soft tofu, and it doesn't look that different from it either.

If you examined a slice of brain under a microscope, it would appear similar to a slice of fatty tissue or your liver. To scientists in the early 1800s, it looked like nothing more than a pale, semi-homogeneous mass.

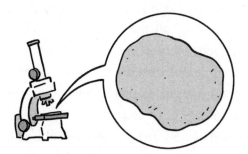

By the 1850s, though, there was a new theory about how living things organized themselves. Better microscopes let scientists get a clearer picture of the tiny structures inside plants and animals. In particular, a new idea came about that helped them see more details in animal samples: staining. Scientists realized that the extra contrast created by adding dyes to animal tissue made it easier to see finer structures. At first, they experimented with different staining agents, including saffron, wine, brandy, and carmine, a reddish liquid made from crushing female cochineal insects and their eggs.*

Using these techniques, scientists realized that living beings are actually made of tiny little blobs called *cells*, and that these cells are the basic units of life. Suddenly, they saw cells everywhere: in our

* Fun fact: carmine is a common food dye used in everyday foods, including strawberry yogurt.

muscles, our skin, our organs, even our bones. The cell theory of life was accepted for every part of our body. Well, every part except the brain.

Something puzzled the budding neuroscientists of the time. If the brain was made up of cells like the rest of the body, what made those cells different from all the other cells of the body? And if those cells were independent of each other, how did they communicate with each other and with the rest of the body?

Examining brain samples using the stains that were available at the time showed nothing but a chaotic mess. Your brain is made up of a dense collection of cells. An average one-millimeter cube taken from its outer layer contains *tens of thousands* of cells. And with every single one of those cells stained, it becomes nearly impossible to tell what's going on, even under a microscope.

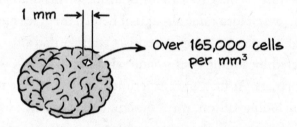

Scientists had one clue, though: they could tell that brain cells had fibers extending out from their cell bodies. They hypothesized that these fibers might be what was connecting the brain cells to each other. But trying to trace which cells were connected together, or even getting a sense of whether those connections had any logic to them, was a pretty much hopeless task.

That's when an Italian scientist named Camillo Golgi came onto the scene.

For years, Golgi worked in secret on a new staining method, never telling anyone about it. When he finally went public, he called it *la reazione nera*, or "the black reaction."

The technique worked something like this:

① Cut out a block of brain tissue.

② Dip the block in potassium dichromate for 2 days.

$K_2Cr_2O_7$

③ Dry it, then dip it in silver nitrate for another 2 days.

$K_2Cr_2O_7$ $AgNO_3$

④ Remove the block, cut out a slice, wash the slice in ethanol, and mount it on a microscope.

The second step, dipping the brain in potassium dichromate, was a common technique at the time to preserve living specimens. It was used to harden the sample, making it less fragile and easier to handle. Golgi's real breakthrough was in the third step: using silver to stain the sample. Silver was a breakthrough not because of what it stained, but because of what it *didn't* stain.

The technique caused dark silver crystals to form inside the cell walls, filling every part of the cell until it was stained black.

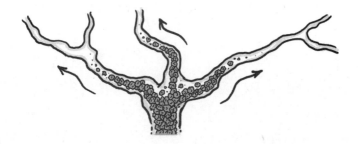

But the amazing thing is that this didn't happen to every cell in the sample. Only a certain type of cell would pick up the silver, and of those only a small percentage (about 3%) would form the crystals.

Before staining, every cell in a brain sample would look like this:

But being able to selectively stain only a few of the cells revealed to Golgi an astounding picture:

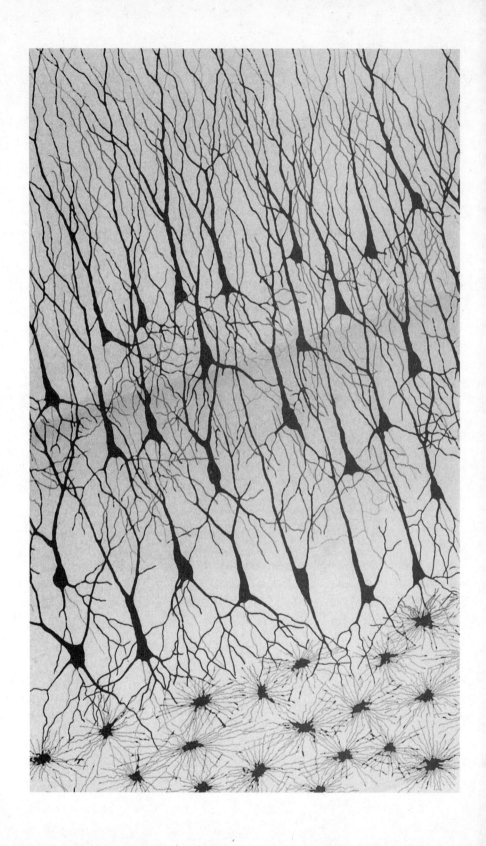

The brain isn't a homogeneous gelatinous goop. It has STRUCTURE.

The images Golgi made confirmed that brain cells had long tendrils that branched and forked like tree roots or tree branches. Here are more images from Golgi's book *Sulla fina anatomia degli organi centrali del sistema nervoso* (1885):

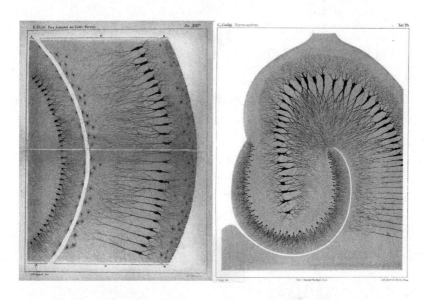

For the first time, we understood that rather than being a chaotic mess, the brain had a certain order to it. Brain cells seemed to have distinctive parts, and those parts seemed to be arranged in a particular way. They appeared to have branches that converged in the cell body, and then a long extension (like a tree trunk) that split off into many more branches. These branches seemed to be doing *something*, as if following a predetermined design. But what was the purpose of that design?

Today we know that the neuron, or brain cell, is the fundamental building block of how the brain works. A typical neuron is made up of three parts: the main cell body called the *soma*, a long projection called the *axon*, and branches of fibers called *dendrites*.

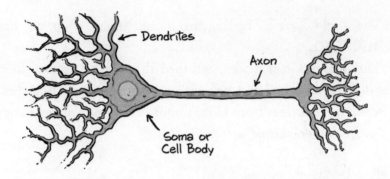

There are over 86 billion neurons in your brain, and they come in many different shapes and sizes. Some are long and thin, others are short and bush-like, and others have odd bulbous shapes. Each of these shapes affects what the neurons do in the brain areas where they are found.

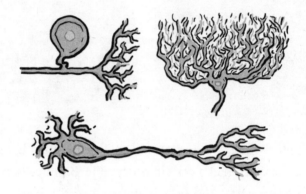

Neurons are not fused together. The dendrites of each neuron reach out to other neurons, but never quite touch. The small gap between them is called a *synapse* and it's through this gap that neurons talk to each other.

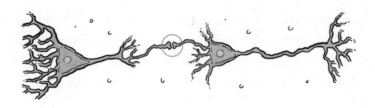

When one neuron sends a signal to another neuron, tiny little pockets at the end of the dendrite release chemicals (called *neurotransmitters*) that travel across the small gap and get picked up by the receiving dendrite (like kids passing notes in school).

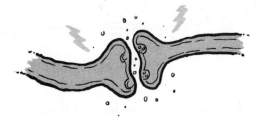

The fact that each neuron is independent is key to what gives the brain its computing power. Each neuron is like a tiny self-contained computer, taking input from one end and creating an output at the other. Neurons "add up" the signals they receive from other neurons through their dendrites.

Some of these inputs can be positive (meaning that they excite the neuron) and others can be negative (meaning that they inhibit the neuron). Overall, if the neuron receives enough positive or excitatory signals, it triggers special gates on the body of the cell to open. These gates allow ions to rush into and out of the cell, thereby creating a surge of electrochemical charge that then travels down the neuron's axon to signal other neurons.

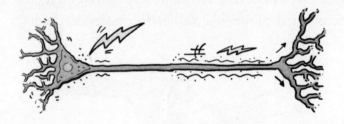

This is how signals traverse the lengthy distances between different parts of the brain and between the brain and the rest of your body. The long axons are covered in a special fatty coating called *myelin,* which insulates individual segments, creating jump points for the electrochemical signal to travel farther and faster.

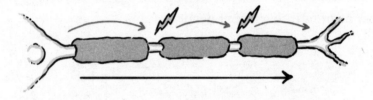

A huge network of neurons with this much interconnection, and arranged in layers for parallel processing, has incredible computational power. Although each individual neuron only does a simple calculation (summing up the signals it gets and outputting a response), the cumulative effect is enough to effectively compute almost any complex mathematical function in just a few steps.

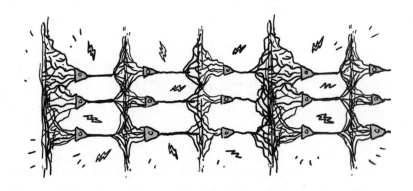

With billions of neurons in your brain, and each neuron typically having over 10,000 inputs from its dendrites, brain networks are very complex and very powerful.

What's more, the network of neurons in your brain is always changing. The neurons are constantly growing and shrinking dendrites, and making and breaking new connections with other neurons along the way.

On top of that, the existing connections between neurons are also changing all the time. Each synapse (the point of connection between two neurons) is continuously being made stronger or weaker depending on how much they are used. This is done by adjusting how much of the neurotransmitters are released at the synapse gap and how sensitive the receiving dendrite is to those chemicals.

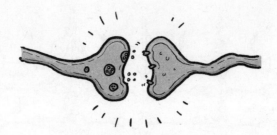

This is how your brain learns and adapts. Imagine a computer whose physical circuits are always expanding and shifting, constantly writhing and pulsing and rearranging themselves like a thick tangle of vines in response to the outside world. That's what your brain is: it's a biochemical computer that, as we'll see in later chapters, has the capacity to reprogram itself.

Rise of the Machines

In the 1940s, scientists started to wonder if we could build machines that worked just like the human brain. In particular, two scientists, Warren McCulloch and Walter Pitts, designed the very first artificial neuron.

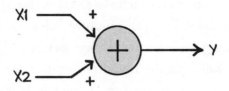

This neuron was a simple mathematical model. Just like a real neuron, it would take in inputs from other neurons and it would make a simple calculation. If the sum of the input signals was more than a preprogrammed threshold value, the neuron would output a positive signal. If the sum wasn't more than the threshold, the neuron would stay silent.

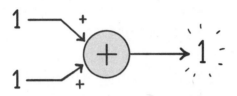

The idea was to simulate what a neuron does. Later versions by other scientists added more elements that you see in real neurons. For example, Frank Rosenblatt, a psychologist, designed the perceptron, which gave different weights to each of the neuron's inputs, and allowed these weights to change.

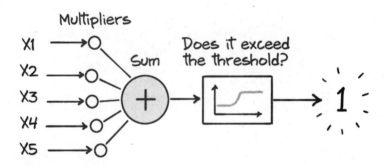

These weights simulate how the points of connections, or synapses, in real neurons can become stronger or weaker.

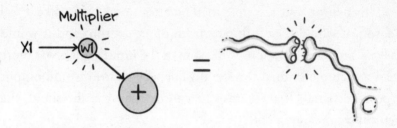

These kinds of artificial brain circuits, or neural networks, were interesting at first. Scientists found that if you connected several of these fake neurons, you could do basic things like recognize simple patterns or make simple decisions. And they found that if you connected more of them and arranged them in layers just like in the human brain, you could do even more complex tasks, such as identify images or make more complicated decisions.

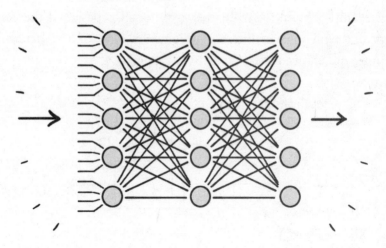

But progress was slow. Adding more neurons to the networks meant there were more connections and weights to keep track of, making it very difficult for traditional computers to crunch the numbers in a quick and efficient way. Luckily, something happened in the computer industry that made artificial neural networks more practical: video games became really popular.

Starting in the 1980s, the demand for more and more realistic video games pushed engineers to make computer chips that could create better and better graphics on-screen. Most computers today have specialized chips that are dedicated solely to making images. These work by doing lots of computations in parallel. Whereas a traditional computer chip can do complex tasks one at a time, graphic computer chips evolved in design to do lots of simple calculations all at the same time. This was necessary because the image on a computer screen has millions of pixels in it, and it'd be very slow for a computer to work on each pixel individually.

These games are great!

In the mid 2000s, researchers at Stanford University's AI lab noticed that graphics cards far exceeded the ability of traditional computer chips to do the kinds of massive parallel computations needed to keep track of large neural networks. Using graphics cards, they could easily simulate and train a neural network with hundreds of millions of weights, cutting the processing time by a factor of 5 to 15.

This discovery turbocharged the development of AIs. Graphics cards became more powerful, and some manufacturers even started to tailor them for simulating neural networks. Today, you can simulate a neural network with a billion artificial neurons using your desktop computer, and the most powerful AIs as of this writing, such as ChatGPT 4, are said to have over a hundred million neurons. It's only a matter of time before we approach the number of real neurons in a human brain (86 billion).

Will They Replace Us?

So, have we accidentally created our future overlords or engineered our own obsolescence? The jury is still out. It's clear that artificial neural networks can be more powerful than the human brain in many ways. This is due to several reasons.

1. *AIs run on faster hardware.* Electronic computer chips are still many times faster than the relatively slow biochemical processes that make real neurons work. As a result, AIs can react, do calculations, and process information quicker than any human.

2. *They have the potential for more brain power.* Right now, the largest AIs don't beat the average brain in terms of the number of neurons each has, but there's technically no limit to how many neurons an AI can have. Future AIs might have hundreds or thousands of times more neurons than we do, making them exponentially more powerful. In terms of thinking ability, it's possible that future AIs will be to us what we currently are to an ant or a worm.

3. *They can be more specialized.* Unlike the human brain, which evolved to do a lot of different things, AIs can be designed and trained to do only one specific task at a time. Imagine if you harnessed the brain power of

the entire human race and focused it on just one problem. In the future, that kind of computational intensity might fit inside your phone or laptop.

It's apparent that, in terms of being outsmarted or outperformed in brain power, we're doomed. As of this writing, AIs have been shown capable of doing complicated tasks that we thought only humans could do, like having a conversation, creating art, making complex decisions, writing essays and stories, and recognizing and manipulating images and videos. Without a doubt, AIs will have an enormous impact on our everyday lives and take over many of the jobs that humans do today. On the plus side, as with any new digital tool, it will also make many of our jobs easier, more efficient, and potentially less prone to human error.

If the fear, however, is whether AIs will take over humans and potentially replace us, then the more important question to ask is "Can AIs have sentience?" That is, can an AI have consciousness or a sense of self?

Can AIs Gain Consciousness?

As we'll cover in chapter 7, consciousness is a really tricky concept to define and to test for. But as we've seen in this chapter, the brain is just a collection of neurons connected to each other. So, in theory, it is possible for an artificial network of neurons to also have consciousness. After all, if it can happen in the human brain, we see no reason why it can't also happen in an artificial network.

However, there are two questions we can ask about this possibility. The first is whether AIs will have consciousness in the same way that we do. We think the answer to this question is "probably not."

The truth is that we don't know how to make a fake human brain. At the end of the day, artificial networks need to be designed, and creating an artificial brain requires more than simply connecting a lot of neurons together. Neural nets have specific structures and different ways that you can arrange the connections and learning loops that will give you different results.

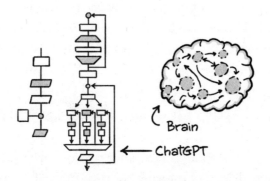

Regardless of whether you're talking about a human brain or AI, there is an architecture to neural networks, and for us this architecture is encoded in our DNA: as we develop (from babies into young adulthood) our neurons are programmed to grow in particular ways and arrange themselves into specific structures (like the cortex, the amygdala, and so on). This architecture is the starting point from which we become who we are.

Right now, we don't understand exactly how to replicate this architecture. We have a rough idea of the different brain areas that exist and how they are generally connected to each other, but the fine details needed to design a perfect copy of the brain are still unknown. As a result, it's unlikely that we could ever make an artificial brain that can truly be human. We may be able to make a neural network that is similar to a human brain, and we can train it to act like a human, but it would technically never *be* human.

The result is that AI may end up developing a form of consciousness that we won't recognize. It's hard for us to imagine a way of being that is different from what it feels like to be human, but it's possible that AIs will develop a kind of sentience that is totally alien to us. As an exercise, try to imagine what it's like to be an octopus, whose "brain" is said to be spread out along its eight tentacles. Or try to picture the "hive mind" of a colony of ants or bees. AIs may develop a sense of self that is just as different from ours, or more so.

The fascinating thing is that since we don't have a clear idea of how to create, detect, or measure consciousness, it could happen in an AI without us knowing it. It might be that some semblance of consciousness already exists in today's AIs. Or it might be that one day we'll create AIs that will suddenly "wake up" and become self-aware.

Will AI Take Over the World?

AIs are capable of giving us tremendous benefits. Artificial networks have been used on medical images since the 1980s, and have reached the point of being as good as human radiologists at detecting breast cancers. In 2023, a neural network was used to control the critical magnetic fields suspending plasma in a Tokamak fusion reactor. AIs have been used to generate more effective vaccines, including against COVID. Already, a neural network called Alpha-Fold has predicted the shapes of over 200 million proteins, which gives hope for similar techniques to discover new drugs and treatments for diseases.

But as with all new transformational technologies, great promise also comes with great peril. AIs might create fundamental changes in our society and economy that happen too fast for many of us to adapt. Some of us might find ourselves replaced by AIs and left without a job (even neuroscientists and cartoonists are in danger of that). And it might be that AI technology grows beyond our ability to control it. Just like it's hard to know what another person is thinking, it's equally impossible to know what an AI is thinking, or to predict what it's going to do. It's totally possible that AIs will develop a sense of self and see us as a threat or as a lesser species to subjugate.

Perhaps we need to slow things down. Some have proposed adding "artificial stupidity" into AI systems by imposing limits on their capacity or speed. Others want to make sure we always include a "kill switch"—a fail-safe that can quickly detect and shut down a misbehaving artificial network. Of course, there's no guarantee that such a fail-safe would always work. In 2023, scientists penned an open letter calling for a pause in the development of more powerful AIs so that appropriate controls could be developed and ethical considerations could be discussed.

Of course, it's also possible for AIs to develop a sense of self and then be nice to us. Maybe training the AI to understand what it means to be human would make it kinder to its makers. One idea is to hardwire a period of development similar to our own childhoods, so that AIs can be enculturated to our human norms and boundaries.

One thing's for sure—if we're ever to understand artificial intelligence, we'll need to continue to explore and better understand how our *own* intelligence works. Applying our collective ingenuity to what is imaginable may one day help us avoid the worst of the unforeseeable.

Chapter 5

WHAT ARE THE LIMITS OF MEMORY?

Sometimes you will never know the value of a moment
until it becomes a memory.

—Dr. Seuss

In 1933, a seven-year-old boy named Henry Molaison went for a ride on his bicycle and his life would never be the same. He fell off his bike and hit his head, leaving him unconscious. When he woke up, it's thought that he had a type of post-traumatic epilepsy, which gave him seizures. His seizures grew from mild spells in which he would lose awareness for several seconds to, by the age of fifteen, severe seizures that caused his entire body to shake and convulse.

His seizures couldn't be easily controlled with the medicines available at the time, and by the age of twenty-seven, two decades after falling off his bike, Henry found he was unable to get a job due to his illness. Having seemingly no other recourse, on August 25, 1953, he underwent a radical, and controversial, surgery.

Henry's doctor, William Beecher Scoville, was an accomplished neurosurgeon, but he was trained at a time when the scientific understanding of epilepsy was still in its infancy. He was at the forefront of something called "psychosurgery," which used surgical interventions, including frontal lobotomies, to address psychiatric problems that didn't have effective treatments.

In a lobotomy, surgical instruments, sometimes even ice picks, were used to pierce the thin bone of the eye socket near the brain, 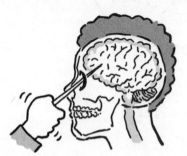 allowing access to the frontal lobes. Once the tool penetrated the brain from below, the lobotomist would swish it back and forth, destroying the brain matter that connected the frontal lobes to the rest of the brain.

At the time, these types of surgeries were celebrated for how easy they were to perform and how fast the results were. Patients with severe psychological problems would suddenly become more passive and docile, and in many cases the surgery was able to stop recurring seizures. One of the technique's most controversial proponents, a doctor named Walter Freeman, did thousands of lobotomies over a ten-year period starting in the 1930s. So, it's understandable how Scoville, who developed and trained in these methods, might think a radical surgery like this was appropriate for Henry's severe form of epilepsy.

In particular, Scoville decided to remove the hippocampus on both sides of Henry's brain. During Scoville's time, doctors had discovered that this portion of the brain seemed to play a significant role in certain types of epilepsy. Given how severe Henry's seizures were, it may have made sense to Scoville to remove all of the hippocampus on *both* sides of the brain, since they didn't know which side the seizures came from. So Scoville drilled holes in Henry's forehead, lifted up the frontal lobes, and not only removed both halves of the hippocampus and part of the amygdala, but also took out part of the frontal lobes. The results were tragic.

HIPPOCAMPUS

The surgery had a profound effect on Henry's brain. While Henry's seizures were more easily controlled afterward, he suffered a very peculiar type of amnesia, or memory loss, that had rarely been seen before. Henry could remember who he was and many of the details of his life leading up to the surgery, but he couldn't make *new* memories.

Every day, he would wake up as if he just had the surgery, and anything that happened to him that day he would soon forget. If he met or talked to anyone, he would chat with them but after several seconds, he would forget who that person was or what they were talking about. He would also forget any new information he read or learned. He could remember his name, his childhood, and everything he learned in school, but at any given moment, he had no idea how he got to where he was or what he'd done earlier that day. He lived almost completely in the present moment, trapped in a sort of time bubble that lasted only about thirty seconds.

Interestingly, he took to solving crossword puzzles, though he could answer clues only about things that happened before 1953 (the date of the surgery). He could also learn physical skills if he practiced enough, like swinging a tennis racket. But, since he had no recollection of practicing any new skill, he'd be continually surprised at how well he could perform it.

Henry Molaison eventually became "Patient H.M." (they used his initials to protect his identity), one of the most important cases in the study of memory. His strange condition was analyzed and tested by doctors, and he became somewhat of a celebrity in the scientific world.

In particular, Henry's case prompted scientists to ask two important questions about how memory works:

Why do we remember some things, but not others?
What exactly are the limits of memory?

How was it possible for Henry to remember some things, especially ones that happened years before, but not others that had happened moments before? This question also applies to our everyday lives. Why is it that we can remember our Social Security number, or the plot of most movies we've ever seen, or specific childhood moments, but not where we left our keys a few minutes ago?

The other question, about the limits of memory, is also profound. Henry couldn't make new memories, and because of his unusual condition, his experience of the world was totally different from ours. Was his life any less complete because of it? Was he stuck as the same person for the rest of his life? How do our memories affect who we are as individuals, and what would happen if we could remember *everything*?

Like a murky recollection itself, our understanding of memory is crisp and clear about some aspects, but completely hazy about others. From what happens at the molecular level when we remember things, to which brain areas enable the formation of new memories, there are still big mysteries. Let's take a stroll down memory lane and explore what we know and don't know about memory. It'll be a trip you won't soon forget.

What Makes Something Memorable?

Memory is one of the most prized treasures of the human brain. Without memory, reading this book would be impossible. Being able to read depends on your having memorized how the different dark squiggles you see represent letters, how those letters are put together to form words and sentences, and how those words and sentences represent concepts and ideas. Without memory, you would just be staring at a mess of undecipherable symbols.

Henry was studied by scientist Brenda Milner. Before Henry, scientists didn't have a firm grasp of how memory worked. One popular theory was that memory worked as a single system of storage and that it was spread evenly across the entire brain. This is why Henry's case was so puzzling. If memory was a function of the entire brain, and his lobotomy removed only certain parts, then what caused his memory impairment?

Milner's breakthrough came after she determined she could teach Henry very specific tasks. Namely, she could get Henry to remember doing things that required motor skills. For example, Milner had Henry trace a shape (like a star) on a piece of paper, but he could only look at his hand and the paper using a mirror. It's a surprisingly tricky task if you try it yourself. But Milner found that Henry got better at it over time, to the point where he could trace the star perfectly. Of course, Henry couldn't remember how or why he got better—he was constantly surprised at how well he could do it.

This odd combination of what Henry could and couldn't remember led Milner to a revolutionary idea: that there are actually different *kinds* of memory and, more important, that each kind of memory can be found in a different part of the brain. For example, maybe the part of the brain that remembers motor skills is in a different section from the one that was removed in Henry's surgery, which is why he could still learn those skills. And maybe the part of the brain that stores *old* memories is separate from the one where recent memories are stored, which is why Henry could remember his childhood, but not what he did more than thirty seconds before.

Milner's insight turned out to be crucial in unlocking a lot of the mysteries of memory. For the first time, there was proof that memory wasn't a spread-out property of the brain, and that the many ways we experience memory are located in different places in the brain.

This is one reason why we can remember some things but not others. In the case of a brain injury to one area, like the one from Henry's surgery, you might lose your ability to store a particular type of memory, but you can still access other kinds.

The other reason why we can remember some things but not others involves how memory flows inside the brain. Most of us don't have brain injuries, and yet we all struggle to remember everything that happens to us. In a normal working brain, what determines what we remember and what we forget?

To understand that, we need to dive a little deeper into our current models of memory, all of which are based on Milner's insight.

How Memory Works

Today, we have a general model of how memory works in the brain. It goes something like this: First, we get information from the world through our senses. We see something, or touch something, or hear it. This information goes straight to a part of the brain called the *sensorimotor cortex,* where it gets processed at a basic level. For example, an image we see might get processed to detect shapes or patterns. While this happens, the image exists as a memory for a short period of time. This is sometimes called sensorimotor memory. It usually lasts only between one millisecond and a second.

If you don't pay attention to this information, it gets erased. Your brain is constantly taking in new data, but most of what we see and hear gets lost because we ignore it. If something does catch your attention, the memory then gets moved from sensorimotor memory to short-term or working memory.

When information is in short-term memory, that's when we're essentially "thinking about it." If you see something and then look away, but you can still picture it in your mind and examine it, that means the image made it to your short-term memory. Or if you hear numbers and you're able to hold them in your head long enough to add or subtract with them, or to write them down on a piece of paper, you're also using your short-term memory.

Short-term memories usually last only around one minute. If you don't keep your focus on them, they also disappear. Scientists think that short-term memory is spread across your frontal lobes. That's the part of your brain behind your forehead where all of your high-level thinking happens. In fact, some scientists don't think there's a difference between remembering something in your short-term memory and the act of thinking. To them, holding something in your short-term memory IS thinking.

But short-term memory has its limits. For example, psychologists in the 1950s started to notice that people couldn't keep more than about seven things straight in their thoughts. They would ask people to compare a certain number of sounds, or to judge how many dots there were in an image pattern, or to try to keep track of a number of objects or digits.

In most cases, people's performance in the tests would get worse if the number of things was more than about seven. These days, scientists have a more nuanced view and think that the limits of short-term memory depend on the complexity of the things you're trying to keep track of. But generally speaking, the limit still translates to about seven simple objects. You can try it yourself: if someone calls out a sequence of random numbers, and you try to memorize them, you may find that after about seven numbers you won't be able to keep up.

After short-term memory comes long-term memory. You can think of long-term memory as the hard drive of your brain. It's where you put things that you want to remember for a long time. Interestingly, our long-term memory isn't in one particular brain area. It's not like your computer, which has a dedicated object (the hard drive) for storing things. In the brain, where a long-term memory is stored depends on what kind of memory it is.

If it's your long-term motor memory (how to swing a golf club, or how to ride a bicycle, for example), then it's stored in the motor areas of your brain, which are located across the top (where you might wear a headband) and control how your muscles move.

If it's your ability to recognize faces or objects, to recognize a song, or to remember the taste of something, then it's in your *perceptual* long-term memory. That memory is stored in your *posterior sensory cortex,* which is near where your hair swirl (called the whorl) is. This is the area of the brain that processes information from your senses.

If the memories are facts, like the capital of Texas (Austin), or the molecular formula for water (H_2O) or how many home runs Babe Ruth hit in his life (714), then these are in your *semantic* memory. Scientists are not sure where semantic memory is stored, but many believe it's spread across the surface of the brain.

3.14159...

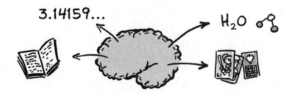

H_2O

If the memory is a flashback, like the memory of being in your childhood home, or remembering what it was like the first time you kissed someone, then it's in your *episodic* memory. This is the most complex of all memories, and scientists think it's the most recent of all memories to evolve in humans. Scientists also think it's related to imagination, and our ability to daydream and project ourselves into the future. Its location is murky, as well, with the leading theory being that it's also spread across the surface of the brain.

Long-term memory can last days, months, or years, up to a lifetime. Even in old age, people can still remember things that happened to them in childhood or places they lived in when they were very little.

The key question now is: How do things get stored in your long-term memory? What determines whether a memory sticks around in your brain, or whether it's forgotten in the dustbin of time?

How Memories Are Etched

Scientists believe that memories move from short-term memory to long-term memory through a process called *encoding*. This process is selective: not every thought we keep in short-term memory gets automatically stored in our long-term memory.

A big part of how things are encoded in long-term memory has to do with repetition. If we want to remember something, say a phone number or a list of items to buy, we usually repeat it to ourselves several times. Or, if you're studying for a test or cramming for an exam, reading and rereading, and forcing yourself to recall something, is a good way to make sure you don't forget it.

Repetition can also happen naturally. If you drive or walk the same way to work or to school every day, that path is probably etched in your brain. Or if you write or type someone's address or a particular password for a website a lot, chances are that you'll remember it for a long time. The same is true for all types of memory: if you practice your golf swing every day, it'll get burned into your motor memory; if you see someone's face all the time, you'll recognize them more easily; if you spend a lot of time in one place, you'll be able to recall a lot of details about it years later.

Repetition requires attention, which is another important component. Attention is how we keep things in short-term memory. If you stop paying attention to something, you'll soon forget it. And remember, once it's gone from short-term memory, it's gone forever (unless you previously committed it to long-term memory).

Interestingly, your brain has mechanisms for "flushing" your short-term memory. For example, a well-studied phenomenon in memory research is known as the "doorway effect." That's when you have something in mind (say, you're looking for your keys), and when you walk into a new room, you completely forget what it is that you went in there for.

To study this, psychologists had people do a simple memory recall test: pick up objects from a table, store them in a box, and

then later try to remember what's in the box. The hitch is that sometimes the subjects would stay in the same room where they picked up the objects, and sometimes they would be asked to go to another room to take the test. The effect was startling: just the act of walking into a different room made the subjects forget a lot of what was in the box. Even walking in and out of the *same room* had the same effect.

Scientists think the doorway effect is your brain's way of keeping your attention where it thinks it needs to be. When you enter a room, it's a new environment—with new potential dangers and opportunities—and your brain is hard-coded to flush whatever you had in short-term memory so that it's ready to take everything in and react.

So, if you want to remember something, make sure you don't get distracted.

Another interesting fact about encoding memories is that it promotes growth. In one famous study in London, scientists analyzed the brains of new and experienced taxi drivers. To be a taxi driver in a city like London (especially in the days before global positioning systems became popular), you had to have extensive knowledge of all the streets and alleys in the city. And London is especially tricky because its streets are curved and arranged in random patterns. The scientists used brain scanning and found that taxi drivers there had a larger hippocampus (the part of the brain associated with memory) than most regular people. The scientists could even track this growth: the longer someone worked as a taxi driver, the bigger their hippocampus.

Where to, mate?

Interestingly, the scientists found that the longer you worked as a taxi driver, the worse you did in other types of memory tests, which suggests maybe there's a trade-off if you overdevelop one type of memory over another.

Repetition and use are one way that memories get moved into your long-term storage. Are there others? It turns out that your brain also has a special trigger for remembering: your feelings.

Emotional Charge

We all have the experience of remembering key moments in our lives. Your first kiss, winning the big game, or the time you suffered a huge embarrassment. Some moments are deeply etched into your memory, and you can recall them as easily as if they happened yesterday. Why does your brain remember these moments so clearly?

For most of those moments, you probably didn't intentionally commit them to memory. You didn't, for instance, cram them into your memory the way you might study for an exam or a test. And these moments usually happened only once in your life, so it's not like they were etched into your memory as a result of repetition.

For these memories, you can thank your amygdala. You might remember the amygdala as your brain's emotional center, where feelings like fear, disgust, and joy are processed. It turns out that this small area of the brain is also involved in helping you remember things.

AMYGDALA

In one set of experiments published in the 1990s, scientists describe studying B.P., a patient who suffered from a rare genetic disease called Urbach-Wiethe. People with Urbach-Wiethe have perfectly normal brains, except for one area: the amygdala. Their amygdala gets damaged because their bodies deposit a lot of calcium in it, turning it essentially into small stones. Because of this, people with the disease don't feel emotions in the same way other people do. They also have trouble recognizing emotions in other people's faces.

In the experiment, scientists told B.P. (whose real name hasn't been revealed since he or she might still be alive) a story that had two parts: the first part was about a boy that walks with his mother to visit his father at work. In the second part, something shocking happens: the boy is involved in a terrible accident, and the subject is shown graphic images of the boy's severe injuries. For most people, the second part of the story had a bigger impact than the first part and they were able to remember more details about it. But for B.P., there was no difference. The first part of the story was just as memorable as the second part, and they couldn't remember more details about one part than the other.

B.P.'s experiment confirmed something that scientists had long suspected: that emotion plays an important role in what we remember.* In later experiments, the scientists scanned the brains of healthy people while they were being shown photos. The photos varied in emotional content, so, for example, one photo might just

* This seems to be partly due to the role of the neurotransmitter noradrenaline, which in a healthy amygdala helps encode emotionally powerful memories.

show a plant, or a clock, while others might show a fearful face, or an angry dog barking. What they found was that people's amygdalae would turn on when they saw emotionally charged photos, and that these were the photos they most remembered after the experience.

In other words, emotions can supercharge the encoding of your memories. These days, scientists think that the amygdala works together with the hippocampus to move memories from your short-term memory to your long-term memory during emotionally heightened situations. This makes sense: if you experience something dramatic that makes you have deep emotions, whether it's good or bad, you probably want to remember it so that you can look for it or avoid it in the future. Of course, this also has negative consequences. If you experience something extremely harmful or traumatic, you're going to have a hard time forgetting it.

The Limits of Memory

The last question we can ask about memory is "What are its limits?" Does your memory have a maximum capacity, or can it keep growing as you have more and more experiences? And how long exactly can a memory last?

There are people who seem to remember everything. For example, Jill Price, a woman in her late fifties from Los Angeles, Califor-

nia, can remember what she did nearly every single day of her life. If you give her a date in the last forty-five years, she can tell you what day of the week it was, what she was doing that day, what she watched on TV that evening, and anything important that happened that day in the world. Her earliest memories are of being a toddler in a crib, and until around age fourteen her memories are not as complete. But after that, her brain started to record everything automatically, which she describes as "nonstop, uncontrollable, and totally exhausting" given the information overload.

Another example is Kim Peek, the man who inspired the movie *Rain Man*. Kim was a savant who could recite from memory over 6,000 books he had read. He could tell you the city that each telephone area code in the U.S. corresponds to, and the zip code of every major city  in America. He was also an expert in over fourteen areas of knowledge, including music, history, geography, and literature.

 Then there is Akira Haraguchi, a retired engineer from Japan who, in 2006, broke the world record for memorizing and reciting the digits of pi. If you're not familiar, pi is the ratio between a circle's width and its circumference. It's an irrational number, which means that its digits go on forever in a random way. For example, here is pi to its 200th digit:

3.14159265358979323846264338327950288419716939937510582097494459230781640628620899862803482534211706798214808651328230664709384460955058223172535940812848111745028410270193852110555964462294895493038196

Haraguchi was neither a prodigy as a kid nor particularly good at math in school. And yet, on October 2006, in a session that took over sixteen hours, Haraguchi was able to recite from memory 100,000 digits of pi in front of witnesses. To date, that record hasn't been broken.

Examples like these are fascinating because they make us wonder what the limits of memory are. The idea of remembering every detail of your life or memorizing 100,000 numbers seems almost impossible compared to our everyday experience. For most of us, keeping track of a grocery list while we go to the store is a challenge.

Is there an upper limit to our memory? To find out, we first need to understand *how* memory is encoded in the brain. We know that memory is organized by type and distributed throughout different areas of the brain, but what is happening at each of those areas? If we know how much information we can store in one neuron, can we sum them all up and get a maximum disk space for the human brain?

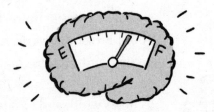

Long-Term Potentiation

The truth is, we don't fully understand how memory is encoded in the brain. We have models that tell us there are different types of memory, and we have an idea of how memory flows inside the brain and which brain areas are involved. But exactly what is happening at the level of neurons, and what kind of language those neurons are using to store information, is unclear.

For example, if you asked how information is stored in a library, the answer is simple: information is encoded in symbols ("words") that are printed on sheets of paper with ink ("books"). Those books are sorted and stored in shelves according to a code (for example, the Dewey decimal system). Or, if you ask how information is stored in your computer, you would say that it's stored as 1's and 0's recorded on a magnetic disk (a hard drive) or in tiny little silicon switches on a microchip (flash drives). For the brain, we don't know what form the information is stored in, and we don't know for sure how that information is recorded in neurons.

The best candidate we have for how information is stored in the brain is something called synaptic potentiation. It's the idea that information is not stored in the neurons themselves, but rather in the *connections* between neurons. It's also a general hypothesis for how the brain changes and learns, a process called *brain plasticity*.

Here's how it works: As we discussed in chapter 4 ("Will an AI Take My Job?"), neurons are connected to each other through long tendrils extending from their bodies. This is how they communicate. A neuron will typically send an electrochemical pulse down its main extension (the axon), and then this signal gets split into smaller branches that then pass the signal to other neurons.

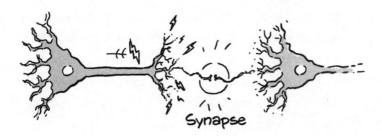

Synapse

What exactly happens at the point of connection between two neurons is important. Scientists call these points *synapses*. As we mentioned before, when a signal arrives at a synapse, it triggers the release of tiny pockets of chemicals called neurotransmitters, which

131

float across the small gap and get picked up by receptors in the receiving neuron.

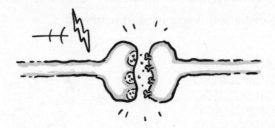

But not all synapses act the same way. Scientists also talk about a synapse's signal transmission strength. For example, signals from a neuron might arrive at a synapse, but the receiving neuron might mostly ignore them and do nothing. This would be called a *weak* synaptic connection, and it happens because the synapse has a low number of neurotransmitter pockets and receptors.

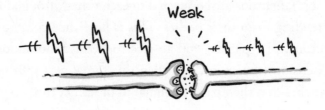

On the other hand, signals might reach a different synapse and trigger in the receiving neuron a big reaction that creates new pulses that continue the signal to the rest of the neuron. This would be considered a *strong* synaptic connection, and it happens when the synapse has a lot of neurotransmitter pockets and receptors, or when the receptors are extra sensitive.

STRONG!

The strength of the connection makes a big difference. If you have a large number of neurons connected together in a network, the strength of those connections is going to determine how signals flow through the network.

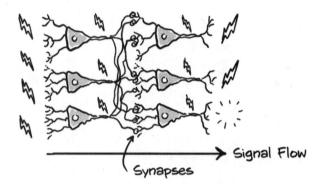

Imagine two identical networks of neurons, A and B, each arranged so that the neurons are connected the same way, but with different values for the strength of the connecting synapses. In one network, some synapses are strong and others weak, and in the other network a different set of synapses are strong and weak.

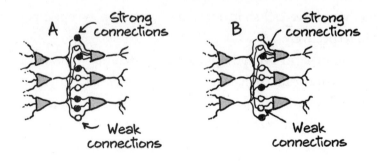

When presented with the same input, these two networks will behave in totally different ways, and their outputs are going to be different. Scientists in the 1950s noticed how changing the strength of the synapses can result in very different outputs, and they started to wonder if this is how memory might work in the brain. The idea is simple: maybe memories are stored in the strength of the synapses that connect neurons together. For example, maybe the output of network A is what makes you remember, say, a dog, while the output of network B is what makes you remember an apple. Depending on which synapses are strong and which ones are weak, you might remember either the dog or the apple.

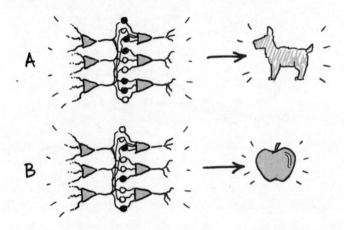

Scientists even found a potential mechanism for how memory gets recorded in your synapses. Your neurons, it turns out, follow a simple principle: the more a synapse gets used, the stronger it gets. This is true over both short periods of time and long periods of time, which we call short-term and long-term synaptic potentiation, respectively.

In short-term synaptic potentiation, sending repeated signals to a synapse can change how calcium ions accumulate in the "sending" neuron. This accumulation makes the release of the neurotransmitter more likely, and recruits more pockets of neurotransmitter to the synapse, increasing its strength.

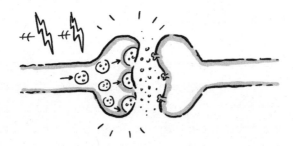

This effect lasts only up to thirty seconds, though. Once the synapse stops firing, the enhancement goes away, which is why scientists think this is a good basis for where short-term memory comes from.

In long-term potentiation, the changes are more permanent. In this case, repeatedly using a synapse has the effect of changing the neuron on the receiving side of the synapse. The receiving neuron undergoes structural changes: new proteins are recruited to build more of the receptors that detect the neurotransmitter, making the synapse more sensitive, and thus stronger. Scientists think this is one way that long-term memories are made.

The basic picture is that the brain stores information through reinforcement. For example, thinking about a sequence of numbers (say, a telephone number) over and over causes the circuits that represent those numbers to be activated over and over again. Your neurons in turn respond by making those connections stronger, reinforcing the numbers into the network. If this goes on long enough, the changes become more permanent, letting you remember the sequence of numbers for a long time.

Interestingly, your brain also seems to have a mechanism for forgetting. It's called long-term depression, and it functions in the opposite manner from long-term potentiation. If you use a synapse less often, it reduces the number of neurotransmitter receptors on the receiving neuron. Your brain does this as a way to keep your memory in check by weakening connections that you don't use as much. It may be why we tend to lose knowledge or skills that we don't access or practice for a long time. In other words, if you don't use it, you might literally lose it.

Estimating Brain Capacity

With this knowledge in mind, we can now try to estimate what the memory capacity of the human brain is. First, let's assume that memory is based on synaptic potentiation, which means that the basic unit of memory is the synapse. How much information can a synapse hold?

We'll assume that the main way to record data in a synapse is in the number of neurotransmitter receptors it has. This is, after all, the best candidate for how long-term memory works: synapses "remember" by changing how many receptors they have on the receiving end of the synapse. This number can be zero (no receptors) up to a certain amount, which is limited by how many receptors fit on the receiving end of a synapse.

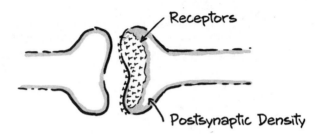

The amount of data a synapse can hold depends on this maximum number of receptors. If the maximum is small, then the amount of information the synapse can hold is also small. But if the maximum is big, the synapse can hold a lot of information. It's like a hard drive: the bigger the hard drive, the more data it can remember.

Considering the width of a receptor (around 12 nanometers), and how big synapses are (typically 250–500 nanometers in diameter), we can estimate that the maximum number of receptors a synapse can hold is around a thousand. This is the equivalent of about 10 bits of data.

Now, the average brain has about 86 billion neurons, and it is estimated that, on average, each neuron is connected to about 10,000 other neurons. This means there are around a quadrillion (1,000,000,000,000,000) synapses in the brain. And if each synapse can hold about 10 bits of data, this puts the estimated memory capacity of the brain at 10 quadrillion bits, or 1,250 terabytes.

To put that number in perspective, the entire text of the *Encyclopaedia Britannica* takes up an estimated 264 megabytes of data, which tells you that the brain can hold the equivalent of around 5 million encyclopedias.

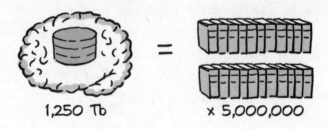

1,250 Tb = x 5,000,000

Improving Your Memory

This estimate, 1,250 terabytes or 5 million encyclopedias, seems amazing. Of course, a lot of the information in your brain goes to running and controlling your body and processing data from your senses. It's just like your computer, where a big portion of the storage space on your hard drive is taken up by the operating system that runs it. It's hard to say how much of the brain's capacity is taken up by basic thinking and information processing.

Is it possible to unlock even more memory power in our brains? The answer might be yes.

One way to expand the limits of your memory is to learn from memory athletes, who compete to set records for remembering things. For example, Akira Haraguchi, who could recall 100,000 digits of pi, is a well-known memory athlete. Others include Munkhshur Narmandakh, the twenty-two-year-old from Mongolia who was the first woman to win the World Memory Championship in 2017. In competition, she was able to commit to memory 6,270 binary digits in thirty minutes. At that same competition, she memorized the exact order of 2,064 playing cards in one hour. Memory athletes go for lightning speed as well: Shijir-Erdene Bat-Enkh, also from Mongolia, memorized the order of a deck of playing cards in a world-record time of twelve seconds in 2018.

As one of their tools, memory athletes use the *method of loci,* or associating things with specific locations. This trick dates back at least as far as ancient Greece, to Simonides of Ceos, who described how he was able to remember the names of all the guests at a banquet by associating each guest with a location within the banquet hall. So, for example, if you wanted to remember a list of things, like "key-woman-kangaroo-hairdryer-dinosaur," you might start by visualizing an image of your house, and then imagine yourself walking through it in a specific path and finding each of those things in a different room. You might tell yourself, "I walked in the front door and saw a key next to the entrance. Then I saw a woman in the living room. Then I saw a kangaroo in the kitchen," and so on. Why is this helpful? The location cues seem to expand your representation of the list, making it more concrete by tying the objects to a place you're already familiar with. It creates a story, helping you add meaning and relationships between the random items in the list.

Interestingly, studies of the hippocampus of mice have shown that certain neurons are keyed to specific places that the mouse has been. For example, one neuron might activate only when the mouse is in the back left corner of the cage, while another may activate only when the mouse is in the front of the cage. These so-called "place" and "grid" cells may give your brain a way to anchor memories and behaviors to specific locations—a literal method of loci.

Recently, scientists have experimented with ways to boost your memory using technology. In one study, scientists used transcranial magnetic stimulation, a technology where a machine sends focused electromagnetic pulses into your brain. The pulses tend to disrupt the activity of neurons, much in the same way a magnet disrupts the information on a credit card or on magnetic tape. This experiment was based on previous studies of people with dementia who developed savant memory skills. The patients with dementia showed a decline in the frontotemporal area near the front of the brain. Hoping to replicate the same effect, scientists aimed electromagnetic pulses at the same brain area in normal subjects. They found that some of the normal subjects were able to do better in certain memory and calculation tasks while their brain was getting pulsed.

But if becoming a memory athlete or disrupting your brain activity with magnets is not in your immediate horizon, there is one thing all of us can do to improve our memory: go to sleep. Scientists have studied the pattern of activation of memory cells in rats and found that rats replayed their memories when they went to sleep. For example, they saw a particular pattern of activity when the rat was walking around and learning a new location, and then they saw the same pattern when the rat was asleep. It's thought that sleep helps consolidate your memories, and that lack of sleep decreases your ability to remember things.

A Final Reminder

There is still a lot we don't know about how memory works. We have a sense of the big picture, and some hypotheses about how it functions at the neuron level, but many of the details still elude us. And while there are numerous examples of people with extraordinary memories, the reality is that we may never know the true limits of our own.

Perhaps a more important question to ask is "Do we *want* to remember everything?" How would our experience of the world change if we had perfect recall? Would we suffer under the weight of our memories, especially the bad ones? Sometimes we may *want* to forget: In cases of severe trauma and post-traumatic stress disorder (PTSD), reducing the intrusiveness of traumatizing memories might be a good thing. Even our everyday interpersonal relationships might degrade if we could not forgive and forget. Forgetting seems to be important enough that evolution has developed unique mechanisms to do it.

There may be one more lesson we can learn from Henry Molaison, the famous Patient H.M. Henry continued to be tested by scientists throughout the rest of his life. He died in 2008 at the age of eighty-two, and his brain was scanned and removed for further anatomical analysis.

Henry was once asked how he felt about all this testing, since he never seemed to become bored of it. Part of it was, of course, that he couldn't remember much from one session to the next. But part of it could have also been his unique Zen-like perspective. To the question of whether he minded being a test subject, he replied: "It's a funny thing—you just live and learn." He then playfully added:

Maybe existing in the present, without the weight of the past, can be a blessing in disguise.

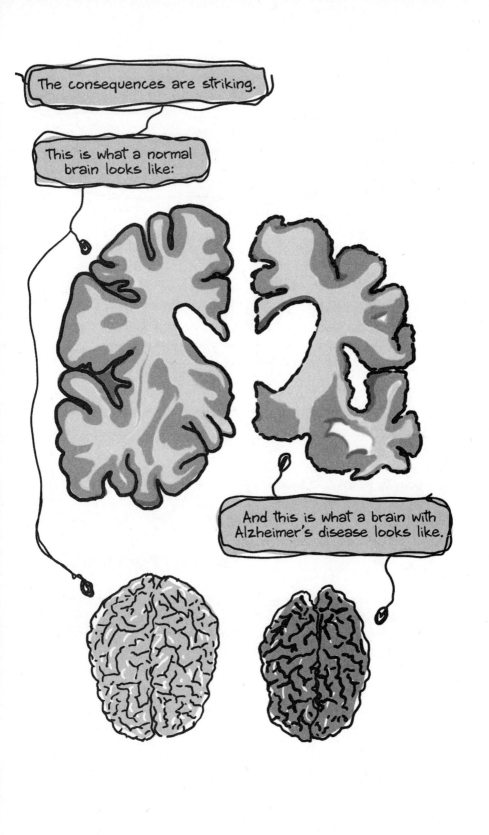

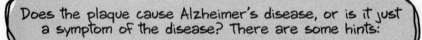

Does the plaque cause Alzheimer's disease, or is it just a symptom of the disease? There are some hints:

Alzheimer's disease seems to be more prevalent in people born with a genetically weaker form of a protein called Apo-E.

ApoE

In normal brains, this protein seems to be responsible for clearing away the plaque associated with Alzheimer's.

And in mice experiments, drugs that increase Apo-E activity seem to reduce the plaque and improve memory.

Alzheimer's disease causes over $300 billion in healthcare costs a year. While you read this, 2 more people were diagnosed with the disease.

For many, time is running out to untangle the mysteries of this common disease.

Chapter 6
WHAT IS ADDICTION?

It is cocaine . . . a seven-per-cent solution. Would you care to try it?

—Sherlock Holmes, in *The Sign of the Four*

These days, it seems we're all addicted to something. Our phones, the latest food craze, the newest TV show. We use the word casually sometimes to mean a certain compulsion to indulge in something, maybe beyond the point where it's reasonable or even healthy. When someone says "I'm addicted to social media" or "I'm a news junkie," it's to express a loss of control, as if something else besides your brain is making you do it.

A lot of things we say we're addicted to fall more in the "guilty pleasures" category. We know it's probably not good to binge on these things, but we somehow end up doing so anyway, usually leading to feelings of guilt and sometimes even shame.

Compulsions in this category aren't that hard to disrupt, though. We commonly do them only when we have the time or the opportunity, which means they fall fairly low on the priority list of our lives.

Of course, addiction also has a more serious medical definition. In recent years, the word has been used to clinically diagnose intense compulsive behaviors like gambling addiction, internet addiction, sex and pornography addiction, and even video game addiction. These addictions are the subject of rigorous medical and neurobiological study because they can have more significant consequences for someone's life. When taken to the extreme, they can disturb a person's mental health and disrupt their social lives and relationships, even cost them their job and financial security. In rare instances, they can even lead to death, as was seen in multiple cases of people who couldn't stop playing video games and died from dehydration, exhaustion, or heart attack.

The word "addiction" also refers to substance addiction, which is the compulsion to consume things like alcohol and drugs. There's a range of how potent this kind of addiction feels: substances like cigarettes and marijuana are addictive but not considered to be as dangerous as "hard" drugs like cocaine and opioids such as heroin and fentanyl. Abusing stronger drugs like these can lead to more dire consequences and completely ruin a person's life, too often ending in imprisonment or in death.

At the same time, it's not the case that everyone who tries drugs, or who gambles or plays video games, becomes addicted.

For example, only a small fraction of people who use opioids become addicted.* This low addiction rate is central to the debate over whether and how much to regulate substances like alcohol, tobacco, or drugs, or activities like gambling. For many people, they are a source of recreation or, in the case of some drugs, part of a doctor-prescribed treatment.

So, how do we make sense of this murky picture of addiction? On the one hand, there seems to be a spectrum of different addictive behaviors, ranging from mild compulsions to serious life-threatening dependencies. On the other hand, not everyone who is exposed to addictive substances or behaviors gets trapped in a cycle of perpetual use.

What does modern neuroscience have to say about this? In this chapter, we'll cover two main questions:

What happens to your brain when you get addicted?

Why do some people get addicted and some do not?

We'll start with what addiction does to the brain. As we'll see, addiction changes the very structure of your brain, reconfiguring your priorities and even, some might say, who you are.

What Is Addiction?

The simplest way to describe addiction is to think of it as a hijacking of the brain's reward system.

As we discussed in chapter 2 ("Why Do We Love?"), the reward system is a set of areas in your brain that are wired together to tell you when something you've done is good. For example, this net-

---

* The addiction rate for heroin is about 25 percent, but the jury is still out for other opioids like fentanyl and carfentanyl.

work triggers your desire to eat both sweet and fatty foods. This is likely because our bodies need energy and, at least for most of human evolutionary history, it was never clear when you would have your next meal. So our brains evolved to feel rewarded when we eat high-calorie foods. There are numerous other examples of our reward system being activated by basic survival needs. It's why a blanket feels so nice on a cold day, or why drinking liquids when you're thirsty is so satisfying. These are all things that contribute to our basic survival, so our brains are programmed to tell us they are good.

The reward system is also activated by more complex needs. As we talked about in the love chapter, we feel good when we meet a person who might be a potential mate, or when we embrace someone we've bonded with. Both of these actions are also important for our survival. The same could also be said about gambling. In this case, evolution might have also found it useful for humans to take small risks now and then. With a big enough population, it would be advantageous to have some incentive to try new things, to venture out beyond the pack, or to hazard going up against a predator that is stalking the village. All it takes is for one individual to succeed, and it benefits the whole group. This could be the reason why, when we take a chance on something despite the fear of a negative outcome, our reward system lights up, giving us a small rush when it works out for us.

Think of any activity in your life that makes you feel great—doing a good day's work; getting exercise; solving a puzzle; spending time with friends; taking the time to rest—and you can probably connect it to some evolutionary advantage that your reward system has evolved to recognize and encourage.

How does the reward system do this? It starts with the small bundle of neurons deep in the middle of your brain called the ventral tegmental area, or VTA.

The VTA receives signals from many different parts of your brain. When it senses that something good has happened, it turns on.

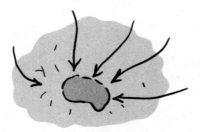

Then the VTA sends signals to other parts of the reward system.

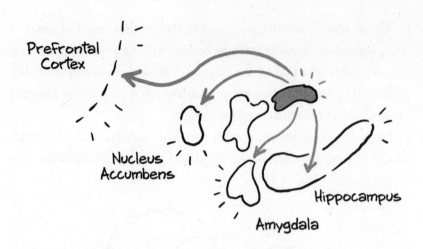

Prefrontal
Cortex

Nucleus
Accumbens

Hippocampus

Amygdala

These areas do different things. For example, the amygdala is where your brain processes emotions. In this case, the VTA sends a signal to the amygdala telling it that whatever activated the VTA is good or pleasurable. Other areas also get turned on by the VTA:

Prefrontal Cortex
(Higher Thinking)
Makes you aware of
what's going on.

Nucleus Accumbens
(Motivation Center)
Makes you want more
of the stimulus.

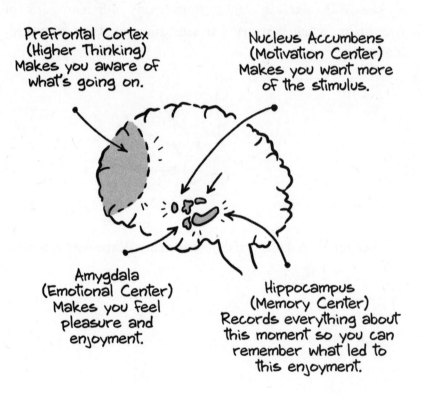

Amygdala
(Emotional Center)
Makes you feel
pleasure and
enjoyment.

Hippocampus
(Memory Center)
Records everything about
this moment so you can
remember what led to
this enjoyment.

The main job of this circuit is to respond to something that your brain thinks is beneficial. It makes you feel good about it, it helps you remember how you got that feeling, and it imprints in you the desire to get more of it.

Normally, this system works well. Over the course of human evolution, it's been fine-tuned to adapt and give us just the right amount of incentive to do things that keep us, and our species, alive. Unfortunately, humans have figured out ways to hack this system, throwing everything out of balance. One way is by creating artificially intense experiences, and another way is by using drugs.

Artificially Intense Experiences

When the reward system evolved, it did so at a very different time in our history. Back then, for example, sugary and fatty foods were not that readily available. You might imagine it was probably rare for our ancestors to find a trove of honey or to hunt down a meaty prey. In the same way, gambling or finding a suitable mate were also probably not very common. As a result, the reward system was calibrated to give us a certain amount of pleasure when we did get to indulge in these things.

These days, however, we live in a very different world. Think about how easy it is to get candy or high-calorie fast food. Or the sheer volume of experiences (shopping, video games, sex, entertainment) that are available to us on the internet with just the click of a button. Gambling has been a part of human culture for thousands of years, but in modern times it has become a mass industry with entire palaces devoted to it (casinos), whole cities built around it (Las Vegas, for one), and all the convenience of gambling websites or apps on your phone. Thanks to human ingenuity, we have distilled, bottled, and made easily accessible almost any feel-good stimulus imaginable.

In other words, we've engineered it so that we can activate our reward system almost anytime, anywhere. And while some might consider it progress to have these pleasures so readily available, it can also have a negative effect. As we'll see later, our reward system was not made for this kind of convenience. Lowering the effort needed to engage our reward system can put our entire brain out of equilibrium, leading to serious addiction.

Drugs

The second way we can hijack our reward system is through chemical substances, namely drugs. Since the brain is a biological machine, introducing chemicals can directly affect how it works. But a key question is how drugs are able to target *specific* areas of your brain. After all, the brain is a giant blob of fatty tissue and billions of neurons. How can a drug pinpoint exactly where your reward system is, and how does the drug get your system addicted?

The answer is that neurons don't use one kind of chemical to talk to each other, they use many different kinds. The brain tends to be organized into different networks, and each network tends to use different types of chemicals for communication. It's part of how your brain organizes itself and keeps its many different parts working somewhat independently. If your brain used only one chemical to communicate, it would run a higher risk of crosstalk between the different networks.

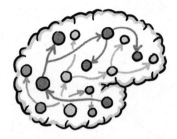

The way that most drugs target the reward system is through dopamine, which is the primary chemical used by neurons in the reward system.

HO — ⌬ — NH₂

Dopamine

Here's how it works: When two areas in the reward system want to talk to each other (for example, when the VTA sends a signal to the nucleus accumbens), they use bundles of long neurons that stretch between them. These neurons transmit signals as bioelectrical pulses that zip across the surface of the neurons' long extensions, or axons.

These neurons might relay the signal directly to neurons in the receiving brain area, or they might relay the signal to other neurons that carry the signal forward.

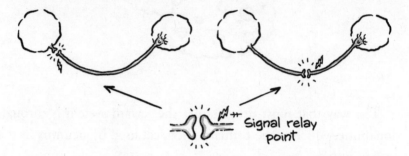

How that signal is relayed is important. When the signal arrives at the interface between the two neurons, it has to jump the small gap that exists between them. It does this by opening small pockets of dopamine in the sending neuron, releasing the dopamine into the gap.

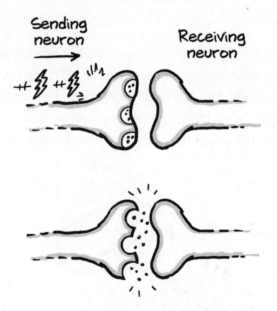

This dopamine floats over the narrow gap to the receiving neuron, where it sticks to special proteins called *receptors*.

156

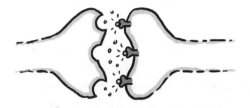

These receptors get activated by the dopamine and cause the receiving neuron to create a new impulse, which continues down the line.

This kind of signal handoff happens all over the brain. It's how most neurons in your head talk to the (on average) tens of thousands of other neurons they're connected to. In total, there are over a quadrillion (1,000,000,000,000,000) of these junctions (remember, they're called synapses) in the brain.

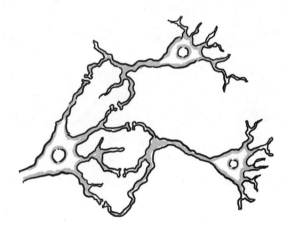

One thing you might have noticed is that this signal transfer happens out in the open, outside the two neurons. This makes the two neurons independent, but it also means that communication between them is exposed. Most drugs that get you high do so by getting into the gaps between neurons in your reward system and messing around with how dopamine is released or recharged. It's a complex system, and each kind of drug does it in a different way. Below is how the major addictive drugs make this hack.

Opioids

Since dopamine is so important to your behavior, your brain has a lot of checks and balances to control how it's released. One of these checks is another chemical called *gamma-aminobutyric acid* (or GABA), which slows down how much dopamine your neurons release in the reward system. Opioids, which include morphine, oxycodone, heroin, and fentanyl, work by suppressing GABA. Without GABA, the dopamine-releasing neurons don't have anything holding them back, and they fill up the junction with dopamine.

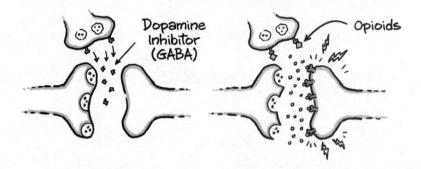

The receiving neuron then sees all this excess dopamine as a strong signal that something good has happened and turns on the rest of the reward system.

Cocaine

Cocaine, which is made from the coca plant, works in a very different way. Normally, after a neuron releases dopamine, that dopamine is reabsorbed into the neuron by proteins called *transporters*. Transporters pump the dopamine back inside the neuron to reset the connection.

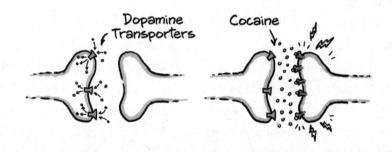

Cocaine works by *blocking* these transporter proteins, preventing dopamine from getting reabsorbed. This causes the dopamine to get trapped in the gap between the neurons, triggering the receiving neuron over and over again, super-exciting your reward system.

Methamphetamine

Methamphetamine, which is usually illicitly taken as crystal meth, is a synthetic chemical first developed in the 1930s as a stimulant (it was given to World War II soldiers to keep them awake). It works using a different hack of the dopamine system: it mimics dopamine, tricking the neuron's transporters into absorbing it into the cell. When the meth accumulates inside the neuron, this confuses the transporter proteins, causing them to pump real dopamine out into the synapse.

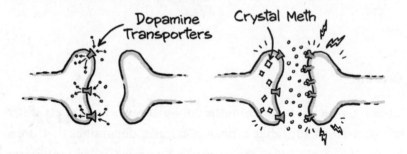

And just as with the other drugs, the receiving neuron interprets the excess dopamine as a signal, turning on the rest of the reward system.

Of course, each drug also has different secondary effects, depending on the chemical and the timing of how it works. For example, cocaine also blocks the reabsorption of another brain chemical, noradrenaline, causing it to flood your system. This is essentially why cocaine gives you an adrenaline high and why you feel jittery or overstimulated when you ingest it. On the other hand, opioids have secondary effects on a different set of brain receptors. Opioids trigger a chain of neurochemical reactions that eventually inhibit different brain areas, including the periaqueductal gray, which is involved in pain modulation and emotional processing. This helps explain why taking opioids has a numbing, or analgesic, effect.

Surprisingly, it's not that easy to get chemicals into your brain. Your nervous system (your brain and your spinal cord) sits in its own volume of liquid, the cerebrospinal fluid, and is isolated from the rest of your body by a set of membranes. The only other bodily system it connects to is the blood system, but there is an added layer of protection in place called the blood-brain barrier or BBB, which covers every artery delivering blood to the nervous system.

Using a series of clever mechanisms, the layer of cells that make up the BBB acts like a filter and lets only certain molecules flow through. For example, it blocks bacteria and large molecules like viruses from going into your brain fluid, and it's very choosy about which nutrients it allows in.

The BBB doesn't even let dopamine through, which is why you can't ingest or consume dopamine directly as a drug or a supplement. If you eat it, or sniff it, or inject it into your veins, it'll just circulate around in your blood. Drugs are special chemicals because they CAN pass through the blood-brain barrier (usually by dissolving through the BBB) and get into your brain fluid.

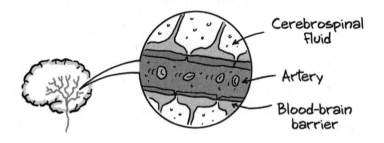

Cerebrospinal Fluid

Artery

Blood-brain barrier

What It Means to Be Addicted

So, what happens when your reward system is hijacked? How does playing a video game or surfing the web create a compulsion to keep doing those things, and how do drugs like cocaine create a substance abuse disorder?

To understand this, we need to understand what dopamine actually does. Because it's involved in the reward system, it's often thought of as the "feel-good" brain chemical. Scientists used to think of dopamine itself as a drug, since its presence is what gives us the high of reward.

This is partly true, but the real case is a lot more complicated. These days, neuroscientists think of dopamine as the "This is important!" brain chemical. Yes, it does give you the feeling of being high, but it does more than that: it restructures your brain. When a neuron detects dopamine, the receptor proteins in the synapse interact with other molecules called G-proteins that can make the receiving neuron more or less excitable. It also changes other mechanisms within the cell. For example, some dopamine receptors affect an enzyme, cyclic adenosine monophosphate (cAMP), that sends messages that cause structural changes in the neuron, like increasing how sensitive other receptor proteins are, or even causing the neuron to make new proteins. In other words, when a neuron receives a dopamine signal, it doesn't just get information about what's happened, it also receives a signal to modify itself.

The reason dopamine neurons do this is to maintain a specific *balance* in your brain, also called homeostasis. This means that your brain chemical levels are stable, and that you have a baseline of awareness and function. It's what you might call "feeling normal."

If something is acting on or stimulating you, your brain will change the sensitivity and disposition of your neurons to balance that stimulus. It's how you get used to things. For example, if you live in a noisy city neighborhood, your brain will eventually adjust to the background noise and start to ignore it. If you didn't, you'd be living in a constant state of awareness of that low-level noise.

In the same way, your brain's reward system balances itself and adjusts to the things that give us pleasure. You can think of it like a seesaw, or a scale balanced on a point:

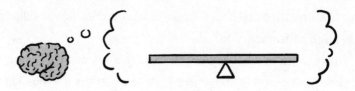

When it's level, we feel normal. When something gives us pleasure, it's like adding a weight to one side. This tilts the reward
system, activating it and causing it to give you a jolt of pleasure.

But something else also happens. The dopamine is also telling
your neurons to change and balance out that stimulus. For example,
it makes your neurons a little less sensitive to dopamine. The effect
is small, so it's like adding a tiny counterweight to the other side.

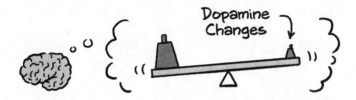

When the stimulus eventually disappears (for example, you
digest the sugar you ate, or the pleasurable experience ends), your
system rebalances itself. Except that now it has a small bias in the
other direction.

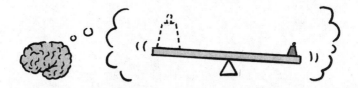

This is how your brain encourages you to seek more of the thing
that gave you pleasure. The slight lean to the other side gives you
the feeling of wanting more of it. Your brain doesn't feel perfectly
balanced anymore, and so you don't feel right until you get more
of the stimulus.

Under normal circumstances, this rebalancing is healthy. Again, it's how your brain encourages you to do things that are important for your survival. But what happens when you hack the system? For example, what if we can activate it too easily with artificially intense experiences, like high-stakes gambling?

Or in the case of drugs, what if we overload the system? Drugs, especially highly addictive ones, have such a direct effect on our dopamine receptors that they act like huge weights (that's why they give such an intense high).

Both of these situations will feel great, but they will also trigger your reward system to rebalance. Your neurons will start to change a lot in response to these outsized stimuli.

Then two things happen. First, you create a *dependence* in your system. For instance, if you deprive your brain of the stimulus (say, you stop gambling or taking drugs), your system now swings hard in the other direction.

This creates a crash, which is known as withdrawal. You feel far from normal, and not in a good way. In the case of a mild addiction, this would be a persistent sense of "not feeling right" or feeling depressed. In the case of strong addictions, like with hard drugs, the feeling is more intense and can even be physical. When someone who is drug-dependent experiences withdrawal, they can become severely ill and suffer convulsions, to the point where their bodies potentially shut down.

The second thing that happens is that your reward system becomes less sensitive to pleasurable input. The thing that gave you pleasure before doesn't have the same effect anymore. That's because your dopamine receptors became harder to activate, and you have fewer of them.

This is called *developing a tolerance* for the stimulus, and it makes it harder to get the same feeling of pleasure the next time that stimulus is received.

This is what makes addiction dangerous. Developing a tolerance for something makes you want to increase the dosage every time. You might seek bigger gambles, or larger amounts of drugs. But increasing the dosages causes your system to change even more, becoming even less sensitive and leading to a cycle of increasing dependence. Eventually, the user may need to take the drug or engage in the addictive behavior not for pleasure, but just to get back to feeling normal, or to avoid being sick.

Addiction alters the structure and function of synapses, but it doesn't stop there. Like a pebble tossed into a pool, these synaptic changes have a ripple effect on the rest of the brain. They can alter the ways the brain communicates with itself and can even alter who you are as a person.

A Different You

Addiction changes the sensitivity of your neurons to dopamine, but it does more than that. Remember that the reward system taps into many areas, including your memory, your emotional center, and your higher executive areas. In changing the sensitivity of your neurons, addiction rewires those areas too. In some ways, addiction doesn't just hack your reward system—it hacks your entire being.

Think back to the experiment we described in chapter 2, where rats were implanted with wires that would electrically stimulate their VTA. At the push of a button, the rats could zap that area in their brain, causing a release of dopamine. In a short time, those rats became addicted to pressing the button, to the point where they ignored basic needs like eating and mating.

The same thing happens to us when we become addicted. Because our reward system becomes unbalanced, other stimuli like friends, or love, or simple pleasures like eating or sleeping are less able to give us pleasure. What we consider good gets rewritten in favor of the addicted behavior or stimulus, drowning out everything else in the addict's life.

Sadly, highly addictive drugs can also have an insidious effect on self-control and your prefrontal cortex. This area is often called the brain's "executive" or boss, because it's thought to be involved in higher-level thinking and deciding.

It's just the sort of place that should help you resist getting addicted, because part of its function is to stop impulsive behaviors. But psychostimulants like cocaine directly reduce how the prefrontal cortex works, making it harder to stop yourself from repeating the addictive behavior. Ironically—or perhaps predictively—the very regions an addicted person needs to fight a compulsion to abuse drugs are some of the very systems attacked by the drug.

Who Gets Addicted?

The last question we can ask is "Who gets addicted?" As we mentioned before, not everyone that tries drugs, or gambling, or video games gets addicted to them. Yet some people seem to be more vulnerable than others. Interestingly, there has been some scientific progress in understanding why.

Experts have determined a wide range of factors that can affect your chances of getting addicted, ranging from your genetics, to where you live and who your friends are.

Seeking Thrills: The first thing that will determine whether you get addicted to something is, of course, whether you try it or not. The one thing that all addictive behaviors or drugs have in common is that they provide a thrill or a rush. Many people try drugs or addictive behaviors in an effort to avoid "normal" or to avoid being bored. In this way, the circumstances in your life (how "boring" your life is, or how "normal" it is) or how much of a thrill seeker you are will impact how likely you are to try addictive behaviors.

YOLO!

Being in unpleasant circumstances, or having underlying mental health challenges, like an anxiety disorder, can also play a role. Studies have found that certain mental health problems can make a person more likely to use drugs if they don't have a way to manage those problems. But just having a mental health challenge doesn't seem to be enough, because plenty of people with them don't use drugs.

The Genetic Lottery: Small tweaks in our individual genetic makeup can cause major differences in how our brains function. We know genetics plays a role in addiction for a couple of reasons. First, there are studies of identical twins (twins who share the same DNA). If one of the twins is a drug user, then it's more likely that the other twin is also a drug user. This is not true of fraternal twins (twins who don't share the same DNA). If your fraternal twin is a drug user, that doesn't mean you are more likely to fall into drug use.

Which genes influence vulnerability to addiction? Research has shown that genetic variations in your dopamine receptors could play a factor. For example, the gene DRD2 encodes the dopamine receptor called D2. Studies have found that people addicted to cocaine have higher variations in the DRD2 gene.

Another gene that might play a factor is called ANKK1, which controls the density of D2 receptors at the synapse. Studies have shown that a variation of the ANKK1 gene is associated with fewer D2 receptors, and with a higher likelihood of cocaine dependence.

Other genetic factors affect how dopamine is processed. For example, some studies have found that changes in the gene that encodes the dopamine transporter DAT (the one that pumps dopamine back into neurons) can lead to more lethal cocaine overdoses because it makes you more sensitive to smaller amounts.

These are only a few of the many genes that we know can affect the part of the brain that is sensitive to addiction. There may be many more. Understanding how these genes change your neurons and how they affect the response to dopamine is of critical importance in understanding and preventing addiction.

For example, there might also be genes that protect you against addiction. While 25 percent of people who try heroin become addicted after using the drug, about 75 percent do not. Those people might be more naturally resistant to it because of their genetic makeup. Recently, scientists have developed genetic tests with the goal of predicting your vulnerability to addiction, based on the presence of certain genes associated with deficiencies in the reward system.

Wrong Place, Wrong Time: Outside factors can also affect addiction. Your environment might be more or less permissive of things like drug use. Peer groups or social situations can also impose social pressures. For example, hanging out in a bar can create an implied expectation to drink alcohol, just as going to a restaurant can create the expectation of having a meal.

Specific places can also reinforce the urge to use drugs or engage in an addictive behavior. This effect, called conditioning, was first described by Russian scientist Ivan Pavlov in the 1920s. Pavlov noticed that his dogs would salivate when they were brought bowls of food. This is a natural reaction, so it's called an *unconditioned response.* But Pavlov saw that the dog would also salivate in response to the noise made by the person bringing the food. The relationship between the sound and the food became conditioned in the dog's brain. Pavlov demonstrated that even neutral sounds, like a bell, could be used to condition the dogs to drool.

In the same way, certain places or stimulation (or even certain people) may be triggering for addicts. In experiments, rats have been shown to keep coming back to the place they were standing when they received a pleasurable electrical stimulation. This compulsion to return to a place where you were stimulated is called *conditioned place preference.*

People who are addicted can experience a similar type of conditioning. The complex stimulation of a particular room or a place can be enough to trigger someone to engage in an addictive behavior. For example, environmental cues like lights, specific music, or certain smells have been shown to increase the rate of smoking in cigarette smokers.

Is it possible to "forget" these associations, or make them weaker? The answer seems to be yes. At least in experiments with animals, it's possible to create conditions that erase these place associations. The most obvious solution is to remove the animal from the environment. Applying this to humans can be hard, though, since triggers can be connected to where the person lives, what they do for a living, or even a loved one.

Personal Motivation: Each person has a different personality and set of social bonds in their lives that make drug use more or less desirable. At the same time, the behaviors of those who are profoundly addicted can harm or even destroy their relationships with family and friends and disconnect them from a sense of purpose in their lives.

Sometimes, the person who is addicted may push away loved ones because they see them as a barrier to continuing their addictive behavior. How afraid the addicted person is of losing those connections can be a big factor in avoiding or ending an addiction.

And just like an environment can promote addictive behavior, being around other addicted persons can create a kind of implied "permission" to continue on that path. They may even support each other's addiction. If, instead, the addicted person is somewhere where sobriety is valued, and the harms are discussed openly, the tendency to use may be reduced. This is the goal of support groups that are a part of many drug recovery programs.

In understanding profound addiction that goes on for years, one often hears about the concept of hitting "rock bottom." How "bottom" is defined is particular to each individual, but in studies of those in recovery this realization—that change is needed—typically takes the form of experiencing direct and extreme harm (for example, the death of a loved one, a health crisis, a personal overdose, or a realization that continued use is unsustainable). In this case, the toll on health, self-care, family, and ability to function can become unbearable, making a change necessary just to survive.

Ignoring the Stigma

To recap, neuroscientists are finding that there are common mechanisms between the different types of addiction. These mechanisms all involve the reward system, but they can vary in addictiveness and potential harm. To add some perspective, here is a published graph of common addictive substances plotted according to those two factors, with an added marker showing how legal they are:

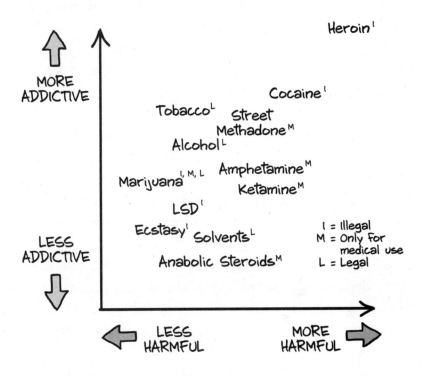

Here, harmfulness is defined by scientists as a combination of physical, dependence, and social risk factors. As you can see, it's not very clear where to draw the line between legal and illegal substances. Marijuana is considered to be less harmful and less addictive than tobacco, and yet it is classified as an illegal substance according to the U.S. federal government (though many states have legalized it).

The important thing to remember is that your brain is both hackable and malleable. Some of us are more or less prone to addiction, so we need to be careful about what we do and the situations we put ourselves in. Something that seems harmless at first can have long-lasting consequences, and what one person finds addictive may not have that effect on another. The problem is that addiction changes you, sometimes to the point where you don't recognize or admit the harm you are doing to yourself.

There's also a social stigma around addiction that complicates things further. According to a recent Pew Research Center survey, nearly half (46%) of all American adults say they have a family member or close friend who is or has been addicted to drugs. This is generally true whether you are a man or a woman, a Republican or a Democrat, or whether you are white, Black, or Hispanic. And yet there is still a tendency to see addiction as a failure. Those with loved ones struggling with a serious addiction often prefer to see the afflicted as they used to be, before the addiction, and resist understanding that addiction changes the structures of the brain and undermines the internal motivations that drive us all.

While the person with an addiction may look the same on the outside, the changes happening inside are profound, and may be cruelly depriving them of the cognitive tools they need to recover. So when dealing with addicts, the best we can do may be to fight through the stigma and meet them where they are and not where we wish them to be.

The Case of The Addicted Detective

Did Sherlock Holmes have a drug problem?

Holmes, the famous Victorian-era fictional detective, was prone to use cocaine.

In those days, cocaine was commonly used as an anesthetic in syrups and tonics. It was even included in the original Coca-Cola formula.

As described by his close friend, John Watson, Holmes exhibits many of the signs of addiction.

But neuroscientists today have wondered if that's really the case, were Holmes real. Let's examine the evidence.

In *The Adventure of the Missing Three-Quarter*, Watson writes that Holmes's cocaine use "threatened to check his remarkable career."

Drug addiction is defined as compulsive use despite negative consequences.

A passage from *A Study in Scarlet* suggests that Holmes suffered from traits associated with bipolar disorder.

Studies have shown that the risk of substance abuse is higher in bipolar patients.

According to Watson, Holmes would use cocaine in his "down" periods between cases, meaning that he kept returning to it.

Drug addiction is also characterized by chronic and relapsing use.

Chapter 7
what is
CONSCIOUSNESS?

I think, therefore I am.

—René Descartes

Hello there! How are you? Are you feeling comfortable right now? Would you like a hot cup of tea? Not a tea drinker? No problem. OK, here is a dumb question: Are you someone else?

Are you by any chance a different person, or are you pretty sure you are yourself? If you are pretty sure you are yourself, then congratulations, you have consciousness.

Don't worry, it's not a disease. In fact, everyone seems to have it. The only problem is that scientists don't know what it is. Or at least they can't agree on a set of words that define it.

You see, the first thing to know about consciousness is that people have a hard time talking about it. Try it for yourself: you are a conscious being. What does that mean

179

to you? Not easy, right? Some of you might think that it means that you are aware of yourself. But what is awareness anyway? Or you could say that consciousness is your sense of self. But what do either of those words ("sense" and "self") actually mean? Some of you might agree with René Descartes, the famous seventeenth-century French philosopher who said, "I think, therefore I am" (in Latin: *cogito ergo sum*). To him, to be conscious is to think. But in a sense, computers also "think" and process information. Does that mean that your laptop is conscious too?

The definition of consciousness has been a topic of debate for centuries, and even more so once neuroscientists jumped into the mix. In a way, what makes it so interesting to study is that humans seem to *know* what it is (we all have a sense of being conscious beings), but nobody seems to know how to describe it.

One of the reasons consciousness is so hard to talk about is that consciousness is a feeling and feelings are generally hard to describe (though as we've seen in previous chapters we've made progress with such feelings as love and hate). But consciousness is much more slippery. There's nothing that we can point to that is the result of having consciousness, except our own unique experiences. It's like trying to describe what it means to see to a person who's been blind their entire life. You only know what it's like to be you, and you don't know what it's like to *not* have a sense of being you.

Can neuroscience help us understand this vague and subjective phenomenon? Fortunately, there are many ways that we can chip away at this, from slicing and dicing brains to studying conjoined twins and using fMRI machines. It's a subject that gets meta in its mysteries, and wary in its awareness of its complexities. Let's see what science can tell us.

Consciousness Has Its Blind Spots

The first thing neuroscience can tell us about consciousness is that there's a lot it doesn't perceive. A good way to understand consciousness is to compare it to our sense of vision. Vision is about more than just seeing things. It's also about *perceiving* them, that is, knowing that something is there in front of us, and recognizing what it is. That's how we build our awareness of the world outside our bodies.

Consciousness works similarly to vision, but instead of giving us awareness of the outside world, it gives us awareness of our *inside* world: the state of our mental processes. Consciousness is the perception of our feelings, our memories, what our senses tell us, and what our brains calculate. And just like our vision, there are things it misses.

VISION CONSCIOUSNESS

If you're not familiar with visual blind spots, here is a quick refresher. Each eye has a blind spot that you're not aware of unless you look for it. For example, cover your left eye, then use your right eye to look at the cross in the figure below:

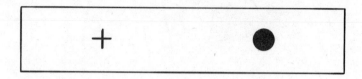

Now move your face slowly toward the page (while still staring at the cross with your right eye). At some distance (around 5 inches), the dot will disappear.

This happens because there is a spot in the back of your eye that doesn't have light sensors—it's where the fibers are that connect your eye to the brain. Curiously, you don't see a hole where the blind spot is because your brain automatically fills it in with what it thinks should be there (in this case, more white background).

This particular blind spot is an anatomical feature of your eye. But there are much more perplexing cases of blind spots. One such case is blindsight.

Blindsight is typically caused by a stroke that damages the primary visual cortex. That's the part of the brain, located at the very back, that processes what we see. Patients who suffer this type of stroke are functionally blind—meaning that they say they can't see anything. But amazingly, if you put something in front of them, they can still react to it. For instance, they can still reach for objects near them or avoid obstacles.

The patients can see things, but they aren't *aware* that they are seeing them. Their brains are taking in the visual information and processing it, but somehow this information isn't communicated to their conscious self. The stroke created a blind spot in their consciousness for visual information. Just like the dot in the figure on the previous page, their view of the world is there, but it disappears from their perception.

What's fascinating about these cases is that they point to the idea that your consciousness isn't aware of everything that's going on in your brain. Basically, our sense of self doesn't "know" everything about ourselves. And this isn't something that happens only because of a stroke or other injury. There are a lot of things happening in normal brains that our consciousnesses are blind to.

Consider the autonomic system. This is the part of your brain and spinal cord that controls your bodily functions, like breathing, keeping your heart beating, and your bowel movements. These functions don't happen automatically: they are regulated by your nervous system, and they respond to what is happening around you. For instance, if you're in a fearful situation, your heart might start beating faster. And yet "we" (our conscious selves) don't control these things (have you tried to tell your heart to stop beating, or your guts to stop churning?). We aren't even aware of what this system is doing most of the time. It's like having a second brain spread out across your body, making sure that you stay alive.

And then there is the subconscious. Psychologists often use the analogy of an iceberg to describe the layers of the human mind. The tip of the iceberg would be our conscious self, or "ego," and it floats on top of the water. This is the part of ourselves that we have access to. But underneath the surface is a huge part that we can't see. This is made up of the preconscious and unconscious mind, or, in popular usage, just the subconscious.

184

Your conscious self

Your deep dark secrets and fears

The iceberg analogy is commonly attributed to Sigmund Freud, but it probably came from an American psychologist, Granville Stanley Hall. It illustrates how most human thought happens outside our conscious self. Underneath the surface are memories (some of them readily recalled, others deeply repressed) and unconscious thoughts and mental processes. Dreams, for example, are thought to be manifestations of this hidden mind. Our biases, tastes, and fears are also developed without our conscious knowledge.

Interestingly, there are ways to access your hidden mind. Subliminal messages can get to your subconscious without your knowing. For example, briefly flashing an image to someone for a few dozen milliseconds (less than the time it takes to blink) allows the person to see the image without being aware they saw it. This can work with words and numbers, and even with emotions.

In one experiment, scientists flashed pictures of faces in front of subjects while scanning their brains. First, the face would show a fearful expression for 33 milliseconds, then it would switch to a neutral expression for 167 milliseconds.

(33 milli-
seconds)

Subjects in this experiment would report that they saw only the neutral face. But underneath, their subconscious saw the fearful face. We know this because the brain scan showed that the subjects' amygdala (the part of the brain that processes fear) lit up when they were subliminally flashed the fearful face.

You can also send a signal to your subconscious by showing each of your eyes a different color-complementary image. If you show your right eye a *red* square and your left eye a *green* square, your brain will think that it's seeing a *yellow* square. That's how the colors combine inside your brain's visual areas.

In one experiment, scientists created hidden messages by drawing images using two colors, and then showing each eye the opposite color-complementary version of the images. If you simultaneously show an image to one eye and the color-opposite image to the other eye, the subconscious part of each half of your brain will see the object, but your conscious self will think it's seeing a blank image.

The scientists created hidden images of faces and houses and showed them to people inside an fMRI machine. They found that subjects would report seeing blank images, but their brain would activate differently depending on whether the hidden image was a face or a house. In both cases, the brain activity was almost exactly the same as when the subjects looked at regular (not hidden) images of faces and houses.

Neuroscientists debate how much subliminal messages can influence our thinking. Some believe the effects of subliminal messages last only a few seconds because they are typically not committed to long-term memory. Others argue that a lot can happen in those few seconds. For example, we might make important decisions in that time, or we might essentially make up our minds about something in those brief moments. Some studies show that subliminal messaging can affect how we vote: in one study done in 2008, scientists flashed the word "RATS" for 33 milliseconds before presenting subjects with a picture of a political candidate (in reality, it was just a photo of a random man in a suit and tie). They found that subjects would rate the person significantly more negatively when "RATS" was flashed than when a more neutral word was used.

Where did the scientists get the idea for this experiment? In the 2000 U.S. presidential election, the George W. Bush campaign released an ad against opposing candidate Al Gore that did just that. As the word "BUREAUCRATS" appeared as part of a larger body of text, the last four letters ("RATS") were briefly highlighted across the entire top half of the screen. The Bush campaign claimed it was an accident.

The main point is that your conscious brain doesn't know what's going on all the time, and that information can slip into your subconscious through your consciousness's many blind spots.

Consciousness Can Be Split

Here's an interesting thought experiment. Imagine if someone opened your head, took a scalpel, cut your brain in half, and then closed your head back up again. You would essentially have two brains. Would that mean there were now two of you?

Would each of you be able to control half your body, and have thoughts and dreams and awareness that were separate from the other half? What would it be like to have two consciousnesses in the same person?

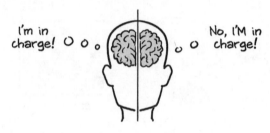

The weird thing is that we sort of know what it would be like, and that's because it actually happened to several people. In the 1940s, this was done to patients who suffered from extreme epilepsy. In the surgery, doctors would open up the skull and cut through the corpus callosum, which is the main bundle of nerves that connects the two halves of your brain's cortex.

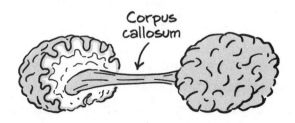

The surgery wasn't always successful (it stopped the epileptic attacks in some patients, but not all of them). And eventually, less drastic treatments and effective medicines came along that made the procedure less common. But for decades, this was a treatment of last resort, and some estimate that it was done to about a hundred people in the U.S., a few of whom are still alive today.

Did these patients end up with two consciousnesses for the slice of one? Sort of.

To understand what happened, you need to know two things about your brain. The first is that the cortex (the outer layers of your brain where all the wrinkles and folds are) seems to be where you do most of your "thinking": it's where you process your feelings, your senses, your use of language, your ability to reason and visualize in three dimensions, and so on. The second thing is that there are two sides to your cortex (the left side and the right side), and they are strangely crossed over. They each get input from and control the side of your body that is opposite to them. So, for example, the *right* side of your brain sees only what is on the *left* side of your field of vision (from either eye), and it controls all your *left* muscles (your left arm, left leg, etc.).

189

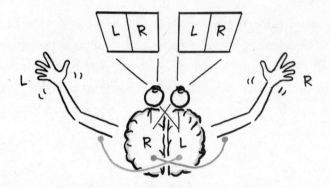

The two sides of your cortex are also different in that they have a few specialized areas. When you talk to or are listening to someone, there's a spot on the side of your left brain that is more active (remember Broca's and Wernicke's areas from chapter 1?). When you are working on a visual puzzle or trying to steer your car around some traffic, the back of your *right* brain tends to be more active.

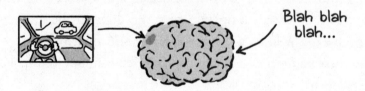

The scientists who studied the split-brain patients came up with some clever ways to see how the surgery had affected their consciousness. For instance, they would ask the patients to sit in a chair in front of a table with a screen. Then, they would flash a word on either the left or the right side of their field of vision and ask them to describe with words what they saw. They would also ask them to draw what they read with their left and right hands. What they found caused a sensation in the world of brain science.

If you flashed a word on the right side of the patient's field of vision, the word would be seen by the left side of their brain.

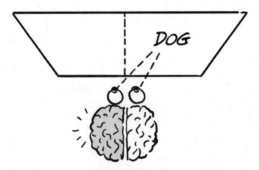

And because this is the side of the brain that is specialized for language, the patient would be able to say what the word was.

But if you flashed a word on the left side of their field of vision, only the right side of the brain would see it. And because this isn't the side of the brain that usually handles language, the patient would report that they didn't see anything. The word never made it to their conscious self. But, bizarrely, they could DRAW what the word represented using their left hand!

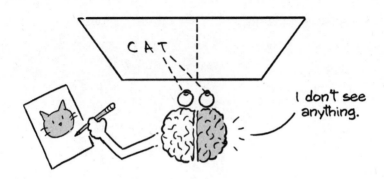

This is because their left hand was controlled by the same side of the brain (the right) that saw the word. In some cases, the patient had no idea why they drew what they drew.

What this means is that, in a way, the patients did have two "brains"! One half-brain could see something in its field of vision and be able to talk about it. The other half-brain could see something on the other side of the field of vision, and be able to draw it, but not describe it verbally. It was almost as if their ability to perceive the world as a whole—and to do things in it—was fractured. One side could do something, and the other side would have no idea about it.

Did this mean that there were two conscious beings inside the patient's head? In general, scientists think the answer is no. For one, none of the patients reported feeling like they had two conflicting personalities inside them. Some reported strange effects right after the surgery, including one woman who said that when she reached for something (say, an item at the grocery store), her two arms would compete with each other for it. She would reach with one arm, and then the other arm would bat the first arm away. (This stopped happening after a few months, though.)

In general, the patients seemed to have a single experience of the world. They were able to lead normal lives, with friendships and family, and they all felt like the same person before and after the surgery.

Scientists think what most likely happened was that the patient's perception and awareness of the world *was* split into two, but their brains were still somehow able to merge those two views into a single internal consciousness. It turns out that, while the corpus callosum is the primary information highway connecting the two sides of the cortex, there are other links between the two halves. One of those is an area in the brain called the thalamus.

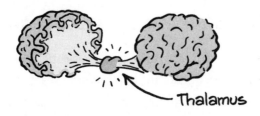

Thalamus

Scientists believe the thalamus plays a big role in our consciousness, acting as the switchboard or hub for all the different areas of the cortex. Perhaps what we experience as consciousness is not only the processing of information that happens in the cortex, but also the integration of that information by the thalamus.

While the main view of scientists is that the split-brain patients retained a single consciousness, scientists also can't rule out a darker possibility: maybe splitting their brain *did* create a separate consciousness, one that is silent and subservient to the dominant personality of the patient. In this case, the repressed consciousness would be trapped and unable to express itself or communicate, yet still be aware of itself.

Let me out of here!

And we may never know if this is true. The group of patients that initially received this rare surgery is getting older, and the treatment is less used now, which means fewer opportunities to study this curious situation.

It's Not the Same as Being Awake

Still with us? Or did you start to fall asleep? Don't worry, being awake is not a prerequisite for being conscious.

By Freud's time, at the beginning of the last century (and as discussed in chapter 4), scientists knew that the brain was composed of neurons, that it's electrochemical in nature, and that many of the areas of the brain had functions that could be mapped (remember the homunculus from chapter 1?). Then in 1949, some researchers made a curious discovery.

A seminal study in consciousness found that stimulating your brainstem seemed to wake up your brain. The brainstem is the nub in the lower part of your brain that attaches to your spinal cord. Inserting a probe into it and giving it a jolt made the brain's electrical activity switch from deep and sleepy waves to short and quick bursts.

Brainstem

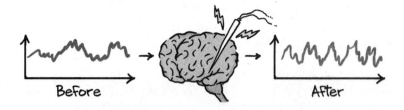

Before → After

Coincidentally, this is exactly what happens when you wake up. In particular, the scientists found this happened when they zapped the part of the brainstem that connects to the thalamus.

But is being awake the same as being conscious?

Wakefulness (being "awake") and consciousness (being "aware") aren't exactly the same thing, but they are related. For example, you can be asleep (which means you're unconscious) and still have consciousness. This is what happens when you dream. When you're dreaming, you are conscious of what's happening in the dream, even if it's weird and you're not in control of the events of the dream. Dreaming happens during REM (rapid eye movement) sleep. In this kind of sleep, the same connection between the brainstem and the thalamus is active, giving you the vague sense of being awake, though clearly the dream is a product of your own mind.

You can also be awake, but not have consciousness. People who have been in an accident or had a stroke can end up in a "vegetative state." In such a condition (which is officially called "unresponsive wakefulness syndrome"), you might show signs of being awake, like opening your eyes or reacting to things, but you exhibit no signs of being "there" or having consciousness.

To test whether consciousness can really be separated from being awake, scientists recently did an experiment in which monkeys were completely knocked out using anesthesia. The monkeys had wires implanted deep in their brains in the central thalamus. When the scientists applied electricity to the wires, the monkeys would show signs of consciousness: they would open their eyes and move their limbs, even though they were heavily anesthetized.

The scientists also checked the monkeys' brain activity using an fMRI machine, and found that when the thalamus was zapped, it turned on a wide range of areas in the monkey's cortex. The scientists also played a series of musical notes for the monkeys when they were anesthetized. Normally, under anesthesia, the monkeys' brains would just ignore these notes. But when the scientists zapped their thalamus, their anesthetized brains actually reacted to the changing musical notes as if they were conscious of them.

This tells us that even in a situation where the brain is completely knocked out, stimulating the thalamus can restore both wakefulness and some degree of awareness. Understanding this could one day help us treat patients who are in a vegetative state.

Consciousness Can Be Shared

Can two people share the same consciousness? To confirm whether the thalamus is really central to consciousness, it would be helpful if there was a way to connect one thalamus to two different brains. And while it's extremely rare, this has actually happened.

Krista and Tatiana Hogan are twins. But they're not just any twins—they share a brain. Their condition is called *craniopagus twinning,* which means they are conjoined twins who are fused together at the skull. Their brain is a melding of two brains with several shared parts, and also the same blood system. Because of this, doctors decided it was too dangerous to separate them surgically at birth.

Each girl has a thalamus, but the two thalami are connected to each other by a bundle of neurons, forming a bridge between them.

Remarkably, they seem to share aspects of the experience of consciousness. For example, they can each see out of their own eyes, but they can also see what the other twin is seeing. If one of them closes her eyes, and you show the other one a toy or a color, the one with the closed eyes can tell what it is. This also works for the sense of touch. If you gently tug the ear of one of the twins,

197

the other one can feel it. Or if something brushes the skin of one of them, the other will notice it.

They can also control each other's limbs. One twin, Tatiana, can control her two arms, and also the arm of her sister. But she can control only one of her legs. Her twin, Krista, can control both of her own legs and one of her sister's legs, but only one of her own arms. The twins even say they can hear each other's thoughts in their brain. They describe it as hearing the other "talk in their heads."

Despite this connection, the twins have unique personalities: Tatiana is described as outgoing and chattier, while Krista tends to be quieter and more relaxed. This is a rare condition—it happens only once every 1.6 million births—and most cases don't survive past twenty-four hours. But, as last described, the Hogan twins went to regular school. They can walk around and they enjoy playing with their dog and talking to friends.

Their amazing experience tells us something about how consciousness works in the brain. Because they each have their own brain cortex, they don't seem to think they are one person. That is, each has her own unique perspective of the world. But, because their thalami are linked, their senses and motor control are intertwined, which creates an overlap in their sense of self. This suggests that the unique feeling of being *You* may be a partnership between deep structures like the thalamus and the areas of the cortex that decode what the thalamus says.

Consciousness Serves a Purpose

Why exactly do we have consciousness? For humans, consciousness seems to be the brain's way of monitoring, perceiving, and "feeling" what it's like to be you. It's similar to the perception of more basic senses, like vision and hearing. Those senses encode an external reality and reconstruct it in your neural circuits. In the same way, consciousness may be doing that with the internal states of the brain and body. It seems to be the subjective, constructed reality of you.

In terms of evolution, we know that the brain developed as a general-purpose problem-solving organ. One key skill for solving problems is the ability to estimate the future consequences of our actions. For example, in trying to come up with a solution for how to reach some fruit on a high branch, you might wonder what would happen if you used a stick, or if you climbed the tree. This gave humans an advantage, and a way to "cheat the system." After all, if you find solutions that don't put you at risk, then you are more likely to survive.

Consciousness might be part of that adaptation. It may be our internal "simulator" that allows us to put ourselves in different scenarios and predict how we would feel about them. If you climb the tree, will you feel good about it, or will it be too much effort? But if you don't get the fruit, will you feel hungry later? Consciousness gives you a sense of self, which you can then put in hypothetical situations.

In other words, we don't actually have to take risky actions—we can imagine them and weigh them against our internal feelings about the outcome of those actions.

It also makes sense from an evolutionary perspective to develop a single conscious experience. After all, we're all individual beings, physically separate from other beings in the world. It wouldn't make sense for our brains to develop two or three or more conscious experiences for one body. It would be too chaotic and disconnected, and it would decrease our chances of survival.

Do other animals have consciousness? If consciousness has evolutionary value, it would make sense for other organisms to have a form of it, too. A dog, for instance, can also benefit from monitoring how it feels about the world around it. But having complex emotions or imagining complicated scenarios may require more brain hardware. As a result, animals might have a much more limited version of consciousness than we do. Or they might not. We'd have to figure out ways to ask those animals about their experience of the world to find out.

Whatever the reason we have consciousness, it's clear that one of its main purposes is to integrate information in the brain. One theory, the global workspace theory of consciousness (GWT), proposes that consciousness comes from the interaction of lots of different brain areas that work together to build a "global workspace." You can think of this workspace as a virtual stage or movie screen that creates a picture of what's going on in your head. Then the higher thinking areas can just watch this movie in order to make decisions about what to do.

According to the theory, your brain operates in two modes: a *local* mode in which information is processed within specific brain areas (what happens "backstage"), and a *global* mode in which information from multiple brain areas is integrated and made available for conscious awareness (what happens "onstage"). The existence of the global workspace is thought to be responsible for the "feeling of being you."

GWT has been influential in shaping our understanding of the neurobiology of consciousness and has been supported by a growing body of evidence from neuroimaging, neuropsychology, and other fields. But, like all theories of consciousness and the nature of our basic neural functions, GWT is still the subject of ongoing research and debate, and there are many open questions and areas of uncertainty that remain to be addressed.

A Conscious Ending

So where does consciousness reside? It's clear there isn't one answer.

The best guess we can make is that certain key areas of the brain (like the thalamus) are necessary for consciousness, but any one area may not be enough to support your sense of self. Consciousness in your brain may be like a Jenga tower: take out one block and the whole thing can fall down.

The fact that consciousness could be distributed across multiple areas shouldn't be surprising. For instance, researchers have found over thirty different locations in the brain that contribute to our sense of vision, and there may be more to be discovered. Each of these areas processes a different set of features, like shape and color. Those features are then coordinated across the brain to provide a single perception of the objects we encounter in the world. Similarly, it could be that multiple brain areas are also required to process the complicated sensation of consciousness.

It could also be the case that you have to *learn* consciousness. Your sense of vision had to become familiar with what objects look like and how they relate to the world outside of us. In the same way, our perception of consciousness might have to learn how to make sense of the activity coming from our own minds. There might be a reason why none of us can remember what it's like to be a baby. It could be that not even babies know what it's like to be babies.

Ultimately, consciousness is the story we tell ourselves about ourselves. It's the sense we use to sort out and interpret our internal, sometimes chaotic brain. And just like our other senses, it can be fooled. This sometimes happens in mental disorders like schizophrenia, where the thalamus and cortex are overconnected. It may explain why schizophrenia patients experience visual and auditory hallucinations and have their own sense of reality.

If there's one thing we've learned about consciousness, it's that it's malleable. As we've seen, it can be split or hidden from view, reduced or even removed. And as we saw with the Hogan twins, it can also be shared.

In their case, the Hogans' brains are physically connected, which gives them access to each other's sensations, feelings, and experiences. Few of us will ever know what it's like to be so directly joined to another consciousness. At the same time, we do get a glimmer of that in our everyday lives.

Every time you interact with someone, you experience a little of what they are feeling and thinking. Every time you read or watch the news, or see a work of art or literature, or view or hear something on the internet, you are also receiving parts of someone else's experience. In a way, this expands your sense of consciousness.

Psychologists often talk about the consciousness of a nation, or the zeitgeist of a time or a culture. Our understanding of consciousness in our brains, and examples like the Hogan twins, tell us that this can be possible.

Consciousness can be shared. For many, that's the purpose of creating art or even writing books like the one you're holding. The more connected we are with each other, the stronger our sense of self as a species will be, and the more likely we'll be to avoid blind spots in how we treat each other.

Chapter 8

WHAT MAKES US HAPPY?

To live the good life . . . be indifferent to what makes no difference.

—Marcus Aurelius

Perhaps no human emotion has been as coveted as happiness. The pursuit of happiness was so important to the authors of America's Declaration of Independence that they defined it as an "inalienable right," which means it's something no one can take away from you. Other countries have also encoded it into their founding documents: the constitutions of Iran, Costa Rica, and Bhutan include happiness, and the supreme law or constitution of China states that its purpose is to "promote the people's happiness." As far back as ancient Greece, happiness was considered to be the ultimate goal. Aristotle is credited with saying "Happiness is the meaning and the purpose of life, the whole aim and end of human existence."

That's a lot of pressure to put on a single concept.

What is this thing that we've been chasing since antiquity, and why do so many of the world's cultures consider it an essential ingredient of life?

Many of the things we discuss in this book—like fear, hate, disgust, or love—are a function of how the brain responds to what's  happening outside of us. We fear, hate, and love other people or other things. But happiness, at least in typical brains, is more about how we interpret what is happening inside ourselves. Scientists define it as a state of mind, which means it's a snapshot of all your mental processes at a given moment, and it's made up of all the rational and irrational feelings about how you are doing. In short, it's your sense of being well (also known as your well-being).

Happiness is a motivating drive. Your brain seeks being in that state because it activates its pleasure centers. Other strong natural drives like sex activate the brain's pleasure centers as well, and certain drugs hijack these centers and strongly stimulate them—but you can't drug your way into a state of happiness. In fact, drug use is often associated with major depression and other mood disorders.

Unlike sex and drugs, happiness depends partly on how you perceive and process what you experience and seems to rely on your personal preferences and values. Just as each of us has a unique consciousness, we also have individual standards of what it means to be happy. In other words, happiness is special because we all find it in different ways, and we all get to choose, to some degree, what it means to achieve it.

Are some people born to be happy? Or are some doomed never to achieve happiness? To find out, scientists have studied identical twins, who share the same genes. In one study of Minnesota twins (actual twins, not the baseball team) they looked at twins who were raised together in the same house, and twins who were raised apart from each other. They also looked at fraternal twins, non-identical siblings born at the same time, as controls. In all, scientists analyzed well-being for four types of twins: identical twins raised together, fraternal twins raised together, identical twins raised apart, and fraternal twins raised apart.

Identical Twins   Fraternal Twins

The study found that identical twins raised apart had a high correlation of well-being. This means that if your identical twin is happy, then there is a good, but not perfect, chance that you will also report being happy, regardless of how you were raised.

This correlation is higher than the one for fraternal twins who were raised together. So if someone had the same upbringing as you and they report being happy, it doesn't mean that you will also be happy.*

This tells us that genetics can play a significant role in determining happiness. Some people are genetically predisposed to be happier than others. But it's far from a perfect correlation. Neither set of twins had a perfect correlation, which is what would happen if genes completely determined your happiness.

Clearly, there is more to happiness than having a sunny disposition, which means there are other factors at play here. But what are those factors? As it turns out, there's a lot that neuroscience can tell us about what affects our well-being, from whether having money makes you happy to what the brain scans of Buddhists reveal about making happy choices. In this chapter, we'll find out if science can help us find contentment in our lives, and whether there can be such a thing as a formula for happiness.

* For the mathematically inclined, the correlation coefficient for the identical twins was 0.48, while the one for the fraternal twins was 0.23.

Finding Fulfillment

In 1943, the American psychologist Abraham Maslow asked a simple question: What motivates people? In asking this, Maslow recognized that humans have a wide range of needs that drive their actions. But he also recognized that not all needs are the same, and that some needs have to be satisfied before others. For example, before you can fulfill loftier needs like having a purpose in life, you first need to take care of more basic needs, like finding food and shelter. Maslow proposed a concept that is now famous in psychology called the Hierarchy of Needs, which is often drawn as a pyramid.

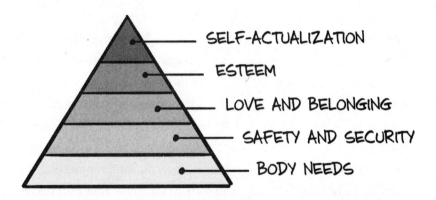

At the base of Maslow's pyramid are essential needs like food, water, air, and shelter. Once you satisfy those, the next level represents safety needs, which include a need for stability, protection, and an emotionally secure environment. The third level represents the need for belonging, which is the need for close relationships with friends and family, and a sense of community. The fourth level is esteem, which includes the need for self-respect and recognition from others. Finally, at the very top is self-actualization, which Maslow described as the ongoing drive to reach one's full potential.

If this sounds to you like he created a video game where you have to unlock levels, you're not wrong! According to this model, we work our way up the pyramid, accruing points that add up to our total happiness.

And at the very top is the ultimate happiness bonus: self-actualization, a state he describes as achieving your true purpose, or becoming what you feel you *must* become. As an analogy for this need, Maslow described the feeling of unhappiness that a musician feels when they can't make music, or that a poet feels when they can't write. This need can be different in each person. For some, it might be the need to be an ideal mother or father, or to reach a certain level of athleticism. Each of us feels an innate potential, he argued, and we can't be truly happy until we fulfill that potential.

Maslow's hierarchy was based on observations he had while working with patients, and there are studies that seem to confirm its validity. In surveys, people often report that the thing they're most stressed-out about is making ends meet. For example, one survey (conducted by the bank Capital One) found that 73 percent of Americans rank their finances as the number-one source of stress in their lives.

In another study, scientists looked at the things people of different socioeconomic classes are emotionally concerned about. They found that people in higher social classes tend to be more concerned with self-focused emotions like contentment and pride and finding a greater sense of amusement. People with lower incomes reported being more concerned with basic emotions such as finding community and gaining support.

Maslow's pyramid paints a picture that makes sense to most psychologists and to most people. It tells us that we can't be truly happy unless we fulfill our basic needs, and that there are different levels of happiness we can achieve.

Can Money Buy Happiness?

If Maslow's hierarchy tells us that achieving happiness is about fulfilling our needs, does that mean that people with more resources have an unfair advantage?

Most of us need money to take care of our basic needs, so money is definitely important. Having money also frees up our time to pursue higher goals, like establishing relationships or working on artistic or athletic endeavors. Money also gives us freedom, and in some sense respect or status in society, both of which are high up in Maslow's pyramid.

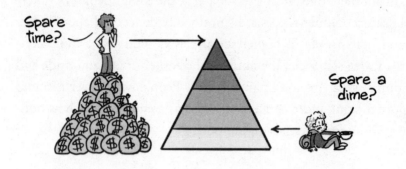

Does this mean that money can buy you happiness? In a way, it can. We like to think of ourselves as rational beings, with higher aspirations than something as seemingly crass as money. But if Maslow was right, money provides the kind of support necessary to meet basic needs and give us access to a more self-actualized life.

To confirm this, scientists have been studying the relationship between income and well-being. In particular, scientists have looked at two forms of happiness: day-to-day happiness and reflective happiness.

Day-to-day happiness is how you feel at any given moment in your life. If someone stopped you suddenly in the middle of the day and asked you "Are you happy?" you might give a wide range of answers depending on what's going on at the moment. But on the whole, your answers would average out to give an indication of how happy you are at any time.

Reflective happiness, on the other hand, is what you might say if someone asked you to sit down and think about your life in general. If you reflect and consider where you are now, what you've been through, and what your prospects are for the future, would you rate yourself as happy or unhappy?

A recent study in 2021 by psychologist Matthew Killingsworth recruited over 33,000 participants, all of them income earners, and asked them to report on both kinds of happiness. To get data on day-to-day happiness, the participants downloaded an app to their phones that would ping them at random times during the day. Each time the phone pinged, the app would ask them "How do you feel right now?" and give them a choice between "Very bad" and "Very good" on a sliding scale. At the beginning of the study, the app also asked them "Overall, how satisfied are you with your life?" and "What is your total annual household income before taxes?"

In total, the study received over 1.7 million day-to-day responses, as well as a wide range of reported income and reflective happiness levels. They found that, for both high and low earners, making more money correlated with both higher day-to-day happiness and greater reflective happiness. In other words, the study found that money *can* buy you happiness (or at least, that money and happiness come hand in hand).

But there was a snag. This study's result didn't match the results of an earlier, very influential 2010 study by some of Killingsworth's colleagues at the University of Pennsylvania. That study, by Daniel Kahneman and Angus Deaton, found that both kinds of happiness increased with income, but day-to-day happiness leveled out at some point. After a certain income, $75,000 per year in this case, Kahneman and Deaton found that people didn't report a significant increase in their moment-to-moment happiness. Their study concluded that money can buy you happiness, but only up to a point: after $75,000, it didn't seem to make a difference in your daily happiness (though it still made a difference in reflective happiness). This study got an enormous amount of media attention, probably because it confirmed what most people want to believe: that there are more important things in life than money.

So, which study is right? Does making more money always make you happier, as Killingsworth found, or does it stop making you happy after $75,000, as Kahneman and Deaton found?

It turned out that both studies were right.

In a rare case of competing researchers working together, Killingsworth collaborated with Kahneman and Deaton to dig deeper into the data. They found that, overall, higher income did increase both types of happiness indefinitely (more money always makes you happier), but *not for everyone*. For certain people, more money stopped making a difference.

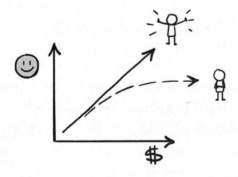

The scientists split the respondents into three groups: people who reported being unhappy, people who reported average happiness, and people who reported being very happy. They found that for people who are unhappy, money stops making a difference after about $100,000/year of income. After that, there is a group of unhappy rich people for whom more money doesn't seem to make them happier. But for people who have average and high happiness, making more money did make them even happier. In fact, for the happy group, having more money *accelerated* how happy they were.

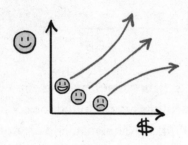

The conclusion is that money does have a positive impact on your happiness. Having more money really can make you happier. But if you have other problems in your life that are making you unhappy, then after about $100,000 no amount of money is going to make you happy. In other words, there are still some problems that money can't solve.

Meaning and Purpose

If genetics don't fully determine our happiness, and if money can't solve all of our problems, where else can we find happiness?

Viktor Frankl was a psychiatrist who asked himself this question, and came to the conclusion that finding meaning and purpose is the ultimate motivating force for humans. Frankl was a Holocaust survivor who, during his time as a prisoner in Nazi concentration camps, saw that

even in the darkest of circumstances, people could still find meaning. He observed that those prisoners that found meaning were more likely to hold on and thrive during and after their experiences in the camps.

So what does it mean to find meaning?

In many ways, Frankl's concept of finding meaning is not that different from Maslow's idea of self-actualization. For Frankl, it's about figuring out what your values are, and then taking action to orient your life in a way that aligns with those values. Or as Maslow puts it, it's about fulfilling the role you feel you were born to fulfill.

According to Frankl, where we find meaning depends on the individual and it can come from many different sources, such as your relationships, your work, or your creative expression. For some, meaning might come from their role as a spouse or a parent. For others, it might come from their career accomplishments or their art.

But while Frankl and Maslow prioritized finding one thing to give your life meaning, most modern psychologists have a broader view of what it means to find happiness. A more recent form of therapy, called "positive psychology," proposes that it's a combination of things that make for a happy life.

A popular model for this is PERMA, which stands for Positive Emotion, Engagement, Relationships, Meaning, and Accomplishments. According to this model, you can find happiness in—

Positive Emotions: Fill your life with activities that make you feel good and bring joy, pleasure, and contentment, such as hobbies or spending time with loved ones.

Engagement: Do things that fully immerse you and challenge you, such as a creative pursuit or a sport. This can result in something called "flow," which is when your mind's attention is completely taken up by an activity. Artists and athletes who report feeling flow experience a sense of enjoyment and accomplishment. Flow can also come from other activities, like playing an instrument or tending to your garden.

Relationships: Seek out positive relationships with family, friends, and your community that build trust and empathy. Studies have found that having fewer, but stronger, friendships makes you happier than having a lot of superficial friendships.

Meaning: Develop a sense of purpose by finding meaning in what you do. Make sure it aligns with your values. If that isn't possible, try doing volunteer work. Studies have found that people who do at least two hours of volunteer work a week are healthier, happier, and live longer.

Accomplishment: Having goals in life is a good thing. Surveys have found that people who set personal and professional goals for themselves are more likely to feel satisfied with their lives.

You might notice two things about this list. First, these are habits that must be cultivated—they all involve being active. And second, they emphasize placing yourself in situations that connect you with an outside focus for your thoughts and actions. That is, one key to finding happiness seems to be to find reasons to get out of your own head.

Some psychologists think this may be a reason why having more money doesn't help the miserably rich. Having more money might come with drawbacks: it puts your focus more on yourself, which leads to social isolation. On the other hand, being poor forces you to seek and build support systems in your community and with friends, which can lead to more fulfilling relationships and a sense of belonging.

Being Happy in the Face of Challenge

All the things we mentioned above can help you find happiness, but what about keeping it? Life doesn't just give us joy—it also gives us challenges and paths that lead to despair. You might experience the death of loved ones, broken relationships, job loss, sickness, accidents, disappointment, and everyday frustrations and stress that can drain your happiness and leave you anxious and depressed. What can you do then?

In 1880, a baby girl born in Tuscumbia, Alabama, would grow up to be one of the most famous figures in American history. Helen Keller was a healthy infant, but at the age of just nineteen months she fell ill with a disease that left her blind and deaf (it is believed that the illness was either meningitis or scarlet fever). As Helen grew older, she became frustrated by her inability to see or hear anything and struggled to communicate with her family members. Her parents, Kate and Colonel Arthur Keller, were determined to help their daughter lead as normal a life as possible, and sought the help of Alexander Graham Bell, who worked with the deaf. Bell suggested that they hire Anne Sullivan as Helen's teacher, and this decision would prove to be life-changing for Helen.

Through Sullivan, Keller learned to say words and to read braille. It was a step out of darkness and silence to find connection with the rest of the world. Keller had a gifted mind. Her educational achievements and support of women's suffrage, worker's rights, and racial equality made Keller a celebrity, meeting with the likes of Mark Twain and John F. Kennedy. She moved through a world she could not directly see or hear, and yet she touched many with her remarkable story. To some, Helen Keller's circumstances might

seem tragic, and she could have given in to the difficulties of her situation. But Keller maintained a positive outlook, saying that a happy life can be found "not in the absence, but in the mastery of hardships." She embraced the challenges that life had given her, and she took joy in meeting them.

Helen Keller faced hardships most of us would never have to face. And yet she was able to find happiness and a sense of purpose in her life. Is this something we can all learn to do?

Half a world away, and 1,800 years before Keller was born, the Roman emperor Marcus Aurelius experienced hardship and challenge in his own way. He ruled during a difficult time of disease and war—his empire experienced a pandemic and fought the Roman-Parthian War and the Germanic wars. We know a lot about him because he maintained a sort of self-help journal called *Meditations*. His entries contained many references to the tenets of the ancient Greek philosophy of Stoicism, which he used to steel himself against the stresses of his position. Today, Stoicism forms the basis or inspiration for many popular forms of psychotherapy, including cognitive behavioral therapy, rational emotive behavior therapy, and even parts of Viktor Frankl's logotherapy.

Happiness: it's all Greek to me.

Stoicism was founded in Athens by Zeno of Citium in the early third century BCE. It's based on the idea that our outlook in life depends on how we respond to the events that happen to us. Stoicism proposes a set of habits for thinking about the world:

1. Look at the world objectively and without emotion.
2. Difficult situations, the actions of others, and even death can only upset us to the degree that we let them.

3. Put others first and work for the greater good, treating everyone fairly along the way.

According to Stoicism, when we experience an event, we tell ourselves a story about that event. That story is based on our biases and expectations, and it's driven by our need to feel that the world is somehow in our control.

But that story may not be true. Or it may be only one of many different ways that we can interpret what happens to us. In either case, the story we use becomes our perception of reality, which can affect our emotions. If that perception is something we didn't anticipate or like, it can trigger a strong negative reaction, which makes us unhappy.

For example, let's say that you lost $100. A non-Stoic might look at this loss with great distress: "That's horrible! I should not

have lost all that money! I really needed that money! I'm so stupid!"
But a Stoic might say, "I lost that money! It's unfortunate, but it's
just money. People lose things. It doesn't mean that I'm stupid. I'll
try to find it. If I can't, I'll just earn some more. If someone finds it,
I'm sure it will benefit them or perhaps they'll return it."

The same event evokes two different responses—one of catastrophe and self-blame, the other of reason and contextualization. A
Stoic might use the habits of Stoicism to rationally assess the loss
and not let it affect their emotions, taking an attitude of benevo-
lence toward whoever may gain from it.

This ability to change your perspective can help you overcome
the things that are making you unhappy.

This basic principle has been used to develop therapies that
are helpful in treating emotional disorders. For example, Frankl's
logotherapy emphasizes dealing with life's challenges by changing
how we respond to them, similar to the Stoic approach. Ultimately,
Frankl argued, we have the freedom to choose the meaning of difficult events and situations, which lets us overcome them.

Rational emotive behavior therapy (REBT), developed in the
early 1960s, also credits Stoicism as a major influence. It's particularly influenced by the Stoic philosophy of thinking logically and
separating events from our emotional reactions to them. In other
words, just because something bad has happened, it doesn't mean

we have to react to it a certain way. The idea is to help patients identify and challenge their assumptions and accept and take responsibility for how they react.

Cognitive behavioral therapy (CBT) is another closely related, popular type of therapy that is based in part on Stoicism. The main idea with CBT is that it's the meaning we assign to events, not the events themselves, that can make us unhappy. CBT aims to make patients more self-aware and realize they have a choice in how events affect them.

An important aspect in many of these therapies is the concept of gratitude or being thankful. Part of changing your perspective and detaching your emotions from events is to essentially "count your blessings" or look on the bright side.

Gratitude can improve your happiness. In one experiment, scientists made subjects feel grateful while an fMRI machine scanned their brains. To do this, scientists asked subjects to read accounts from Holocaust survivors describing moments in which they had been helped. Sometimes, the help was very important, but required little effort, such as the story of a local baker who would leave unsold bread in an alley for Jews to eat. Other times, the help would come at great risk, such as the story of a concentration camp prisoner who risked her life to steal food for a fellow prisoner who was sick. Scientists would then ask the test subjects to imagine what it would be like to be the person who was helped, and to think about how they would feel if they were in the same situation.

The study found that feeling gratitude had the effect of increasing activity in the anterior cingulate cortex and medial prefrontal cortex, two brain areas that are important for maintaining your emotional balance.

As a result of studies like these, psychologists recommend making gratitude part of your regular routine, such as recording the things or events that you are grateful for every day.

The Curse of Choice

Sometimes, despite our best efforts to gain perspective, we still struggle with feelings of unhappiness. In that case, science recommends taking a look at how your life is set up. And, in particular, how choice plays into your life.

Anyone who has shopped online probably understands the "decision paralysis" that comes from being offered too many choices. Shop for any item, and you'll likely be faced with dozens of options, all with slightly different features. Having so many choices sounds like a good thing, but it actually works against your happiness.

Studies have shown that having too many options can lead to frustration, stress, and even regret—the feeling that you made the wrong choice. In a now-classic study from 2000, scientists set up a tasting booth at a fancy supermarket in California. Sometimes the display would feature six kinds of gourmet jams, and sometimes it would feature twenty-four. The scientists found that more people would stop at the booth when it featured twenty-four varieties, which means customers were attracted to having a lot of choices—but they were also *less* likely to buy any of the jams. Only 3 percent of people would purchase a jam. Now compare that to the amount of people who bought jam when there were only six choices—30 percent—and it's clear that having lots of choices was attractive to people, but it didn't help them make a decision.

The scientists then tried a similar experiment with chocolates. They presented subjects with a tray that contained either six or thirty different flavors of Godiva chocolates, and they were asked to choose one flavor. Once they had made their choice, the subjects would fill out a survey asking them to rate their experience in terms of enjoyability, difficulty, and frustration. Then the subjects were allowed to eat the chocolate they picked, at which point they filled out another survey asking if they enjoyed it.

Overall, people liked having lots of options to choose from (they preferred the tray with thirty options). But they found the process of choosing among that many options more difficult and frustrating.

When asked if they liked the chocolate they picked, subjects who had to choose from thirty options were more likely to say no. The people with only six options seemed to enjoy their chocolate more and feel that they made the right choice. In other words, having a lot of choices sounds great, but it's actually more frustrating and it makes you like what you choose less.

Fear of Missing Out (aka FOMO)

Having too many choices also makes you compare yourself more to others, and you're more likely to second-guess yourself. This is called "social comparison," where you not only think you made the wrong choice, but you think someone else made a better one.

And social media has only made this worse. Since most people only share photos of themselves having a good time or living wonderful moments, this can give you the impression that everyone else is living a better life than you.

Sadly, science has shown that envy is a real thing, and that it can affect your happiness.

In one study, scientists looked at how people's happiness changed depending on how much income they thought their neighbors were making. The scientists used data from several surveys done in the U.S. since the 1950s in which people are asked these questions:

- How happy are you?
- What is your income?
- How do you think your income compares to others?

The study found that people's happiness increased the higher their income. But their happiness decreased the lower they thought their income was compared to others'. In other words, making more money makes you happy, but knowing that other people are making more money than you makes you unhappy.

You might be familiar with this feeling if you ever found out you were making less money for performing the same work as a co-worker.

The scientists think this helps explain why in advanced countries like the U.S., overall income has been increasing for decades, but individual happiness has stayed the same over the years. Even though people are making more money in the U.S., the increasing income inequality in the country is making people feel less happy.

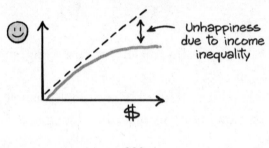

Reducing Your Choices

Of course, no one wants to give up having choices in their lives. But sometimes not having *any* choice can make us happier.

Medical decisions are among the most stressful choices people can be asked to make, especially if it's for a loved one who can't decide for themselves, like a child or someone in a coma. These are complicated decisions that can have a deep impact on the patient's health.

In one study, scientists looked at how parents felt after having to decide whether to end life support for their baby. People usually say they want more control over their health decisions, but here, they found that the parents felt worse when they had to make the decision themselves rather than have doctors make it for them. Why? In follow-up interviews three months later, many of the parents blamed themselves for the outcome, even when it was unavoidable.

This contradicts the idea that more choices make us happy, and it also runs counter to current healthcare practices. Healthcare today emphasizes putting more control over medical decisions in the hands of patients. But having responsibility for these kinds of choices could be making people unhappier. Taking on that particular kind of burden affects how you cope with the consequences and can lead to blaming yourself if the outcome is bad.

People like the idea of having choice but living with difficult choices can be hard. In some cases, learning to trust others to make complicated decisions for you can make you less unhappy.

Is there a way to reduce the everyday unhappiness that comes from having too many choices? In this, there is something we can learn from Buddhist monks.

Buddhist monks aspire to live a life of simplicity and minimalism, with fewer possessions compared to the average person. Buddhist teachings emphasize the importance of reducing attachment to material things and desires in order to achieve inner peace and happiness.

Reducing choices lets Buddhist monks avoid the stress and anxiety that can come with making complex decisions. This can lead to a sense of clarity and calmness, with fewer feelings of indecision and worry. Their minimalist lifestyle also reduces feelings of social comparison that can be a source of unhappiness.

Most of us would prefer not to become Buddhist monks, though. In that case, is there an optimal number of choices to have in life? Research suggests the answer is yes.

A study performed recently looked at this question while scanning the brains of test subjects with an fMRI machine. The subjects were asked to browse sets of six, twelve, or twenty-four landscape images to be printed onto a gift item for themselves. Then the subjects were asked to choose one of them. The scientists found that when the subjects were just browsing the images, their brain activity was the same whether there were six, twelve, or twenty-four options to look at. But when they were asked to make a choice, their brain activity changed.

When they had to choose from twelve images, areas of the brain related to value (in this case, the striatum and anterior cingulate cortex) were highly activated. But when they had to choose between six and twenty-four images, this brain activity was less. This matched what the subjects reported verbally: twelve items seemed to be the Goldilocks or "just right" number of options.

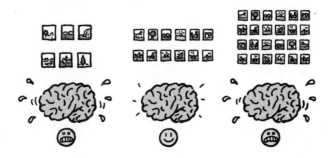

So, the next time you're faced with an array of options, remember that sometimes, less really is more. Design your life so that your options are large enough that you aren't forced to live like a Buddhist monk, but small enough that you don't have all the frustration, second-guessing, and self-blame that come from having too many choices. And if having too many choices is unavoidable, seek the opinion of others. At least you'll be happier with the choices you make.

Depression

Finally, it's hard to talk about happiness without talking about its seeming opposite, depression.

A PRIMER ON DEPRESSION

According to psychiatrists, major depression is characterized by a loss of joy, a lack of energy, changes in sleeping and eating habits, and feelings of sadness.

In severe cases, it also involves thoughts of self-harm or suicide.

Depression is estimated to affect 5% of all adults worldwide, and is one of the leading causes of disability in working-age adults.

Just as happiness and contentment are persistent brain states, major depression can be, too—a dark mirror image of happiness.

There's no clear consensus about what causes major depression, but the leading theories point to imbalances in brain chemicals, genetics, or traumatic stress responses.

Serotonin and noradrenaline are two brain chemicals that might be involved.

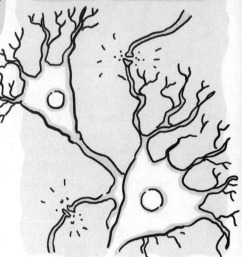

Serotonin is used in mood and sleep regulation and noradrenaline is used in your body's "fight or flight" response.

Imbalances in how these chemicals are produced, or get reabsorbed into your neurons, could lead to areas of your brain not working properly.

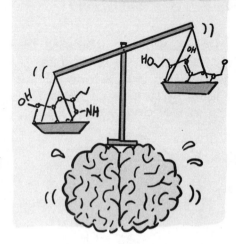

Another possible cause of depression is how we respond to stress.

When you're stressed, your body releases cortisol, a hormone that protects your body from the effects of stress.

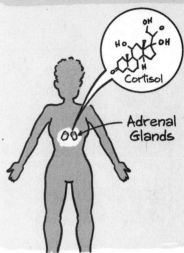

But people who've suffered from trauma (highly stressful or persistently stressful situations) can develop an overactive cortisol response.

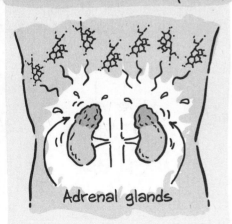

In depression, cortisol stays elevated, which can lead to changes in the structure and function of the brain areas that regulate mood.

Many current antidepressants target low serotonin levels, but most take weeks to work, and for some people, they simply don't work.

The truth is that scientists aren't sure why certain drugs work, they just do.

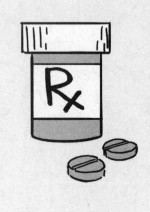

One hypothesis is that antidepressants help your brain build new neurons.

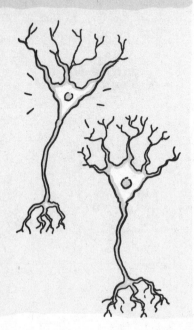

The drugs might be helping your brain rewire itself to make up for disrupted circuits.

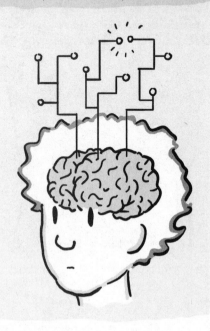

While there are recreational drugs that can create euphoria and pleasure, there is as yet no magic pill for lasting happiness.

The good news is that combining medication with psychiatric therapy can be significantly more effective than medication alone.

The important thing is to not be afraid to seek help. There's still a stigma associated with mental illness that we do not apply to other illnesses. Yet, mental illness is no more anyone's fault than catching the flu.

So, there is hope. Existing treatments and medications can help, especially in combination with other strategies for finding happiness.

Is There a Formula for Happiness?

To recap, there are many things that can affect your happiness. While there's no one solution or special trick to achieve happiness, there are a lot of things that science has taught us about improving our chances of reaching a state of happiness or contentment:

1. Fulfilling your needs is important. Acknowledge you are a human being and that you have basic needs that have to be taken care of before you can hope to be happy.

2. Money is important. Understand that money does help you be happier. But realize that money is a means, not the end itself. There are things money can't solve and having money doesn't guarantee you will be happy.

3. Find meaning and purpose. Take a step back and figure out what is important to you and how you can contribute to society. You are more likely to find happiness when your life is aligned with those values.

4. Nurture positive relationships. Relationships are a big source of meaning and support in our lives. Research has shown that having deep relationships makes us happier and helps us live longer.

5. Take charge of your perspective. Like Helen Keller and the Stoics, remember that you have a choice in how you view the things that happen to you. Think critically, not emotionally, about your situation and practice gratitude and generosity as much as you can.

6. Don't overthink it. Sometimes we're unhappy when faced with too many choices. Simplify your life and try not to get lost in the abyss of indecision. Trust others, and don't second-guess yourself when you take a step forward.

 7. Avoid social comparisons. Don't fall into the trap of comparing yourself to others. Realize that it's healthier to focus your attention on what you have, and where you have room to grow.

 8. Seek help if you have depression. There is a lot that is out of our control when it comes to our mental health. Don't be afraid to get medical help or counseling if you need it.

In addition, there are physical habits that science can point to as helpful for a happy life:

 - Reduce stress: Stress has real physical effects on your body and your brain. If it's self-imposed, ask yourself if the stress you are feeling is worth the negative impact it's having on your life.

 - Exercise: Studies have shown that regular exercise promotes the release of endorphin hormones, which put you in a good mood. And it may even promote the growth of new brain cells.

 - Sleep: Sleep is very important for your mental health. Studies have shown that lack of sleep is related to depression and other mood disorders.

- Meditation: Studies have shown that meditation reduces stress and improves feelings of happiness. Meditation has even been reported to make your brain bigger in the areas that control your positive mood.

Of course, many of the practices or conditions we describe here may depend on the situation or position of privilege you find yourself in. But to the extent that you have control, it is worth examining these areas to see if any of them can be improved or are a source of unhappiness in your life.

Interestingly, worrying about being happy can make you *un*happy. Research has shown that in developed nations with high overall levels of happiness, such as Denmark and Finland, people report feeling unhappy because they feel societal pressure to be happy. Just like the idea of social comparisons over money or making bad choices, it's possible that we feel unhappy with ourselves if we perceive we're not achieving a socially acceptable level of happiness.

So, if there *is* a formula for happiness, it might just be to take it easy on yourself. No pressure!

Chapter 9
DO WE HAVE FREE WILL?

I have noticed even people who claim everything is predestined, and that we can do nothing to change it, look before they cross the road.

—Stephen Hawking

In 1964, under the afternoon sun of Córdoba, Spain, a scientist, José Delgado, stood inside a bullring with a small remote control in his hand.

The assembled crowd watched in silence as a bull was released into the arena, and they held their breath as the bull charged toward the scientist with the fierce determination of bulls that are trained to fight in the ring. But as the animal bore down on Delgado, he pressed a button on the remote, sending a wireless signal to a receiver attached to the bull's head.

The receiver was connected to thin wires that sent electrical impulses into the bull's brain. The wires were inserted through the skull deep into the middle of the brain, stimulating two regions, the caudate nucleus and the thalamus. When Delgado pressed the button, the bull came to an abrupt stop,  and the crowd erupted in applause and amazement. Delgado then left the ring and stood behind a barrier. When he let go of the button, the bull continued his charge and rammed the barrier.

The "bull experiment" captured public attention and was featured in newspapers across the world. This flamboyant demonstration inspired a generation of scientists, and eventually the concept of deep brain stimulation led to treatments for disorders such as epilepsy and Parkinson's disease.

Delgado's experiment also did something else: it began a conversation in neuroscience about the nature of free will.

Every day, we make dozens, if not hundreds, of decisions as we go about our lives. Maybe they are small decisions, like whether to have an apple or a banana for a snack. Or maybe they are important, consequential decisions, like whether to stay in a job or look for another one. In most cases, we make a decision and we live comforted by the feeling that we are in control of our lives. That's because we believe that we could have made different choices if we had wanted to. That is, we could have chosen a banana instead of an apple or quit instead of staying with the job we had. That idea, that we had the option to make a different choice, is what philosophers and scientists call free will.

Free will is the sense that nobody controls our actions, and that our decisions are not predetermined or known in advance. But is that really the case? Do we actually have that freedom of choice, or are we like Delgado's bull, acting at the behest of larger forces beyond our control?

In this chapter, we'll ask three questions related to free will and the brain:

Are we really in control of our actions?
Can someone hijack our free will?
Is our brain predictable?

As we'll see, free will lies at the strange and perplexing intersection of philosophy, quantum physics, and . . . zombies. Of course, whether you read the rest of the chapter is entirely up to you.

Or is it?

Is it a banana day or an apple day?

Are We in Control of Our Actions?

Across nature, there are many examples of creatures doing things they could not have learned on their own. These are freakishly specific behaviors that are extremely complex.

One example is web weaving in spiders. One kind of spider, the orb-weaver, constructs some of the most beautiful webs on the planet, and yet mother orb-weavers die after mating and laying their eggs. This means that baby orb-weavers are left on their own to survive—with no way to learn from other spiders how to make a web. And yet when they're adults, the spiders somehow know how to do it exactly the same way as their mothers did. The spiders adapt to different circumstances—adjusting for different locations, for example—but they always follow the same sequence of steps: First, they lay down a series of lines (called a proto-web), letting the wind push them around to explore their surroundings.

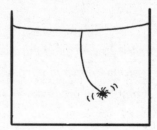

Then they locate the center of the proto-web and start to construct spiderweb lines that extend to all sides, taking down the old proto-web by eating it.

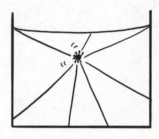

Finally, they crawl along the web in circles to create the familiar spiral of orb webs.

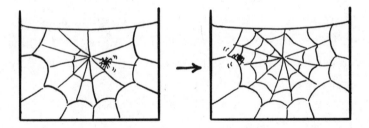

What's fascinating is that each spider knows how to do all of this on its own, without anyone showing it how. That means that the impulse to make the web, all the details of what to do and how to do it, are encoded in the spider's tiny brain.

Birds are another prime example of animals with innate behaviors. Some species of birds, even those raised in isolation who never see other birds, know how and when to build nests. Similar to how spiders adapt to their location, birds might gather and use different materials, but the idea to build a nest, and the skills needed to make one, seem to be preprogrammed in them.

This points to the possibility that many of the behaviors *we* exhibit could also be preprogrammed in our brains. They might be hard-coded in our DNA and how our brains are wired together. If this is true, it suggests we're not in complete control of our actions.

Humans have a wide range of innate behaviors. Some of them are simple—for example, yawning is an innate behavior seen in most vertebrates. And yet nobody learns how to yawn. People who are blind and deaf and have never seen or heard someone yawn can do it. Fetuses are even known to yawn in the womb.

Other innate infant abilities include grasping, suckling, and crying. There's also sneezing, yelling when we're startled, and going to sleep when we're tired. There are a lot of things humans do without explicitly choosing to do them.

Gesundheit.

What about more complex behaviors? As seen in other chapters, complex emotions like loving and hating are common in all humans. Each of these has aspects that don't seem to need learning, which could mean they are preloaded in our brains.

Another example is our innate need for shelter. When we choose to buy a house, or set up a tent, or look for a place to sleep, we are likely following some primal instinct written in our programming, just like the spiders and birds. If everything that goes into building a complex spiderweb can be written in the 100,000 brain cells of a spider's tiny brain, it's not hard to imagine that many of our complex behaviors are also programmed in the 86 billion neurons in our brain.

The point is that a large part of what we do, we don't really choose to do. It's predetermined in our genes and how our brain areas are wired together.

Of course, our DNA doesn't have absolute control over us. We can stifle a yawn or suppress a cry. And we can certainly stop ourselves from buying a house or putting up a tent. But not always.

When we think about free will, drugs naturally come to mind. In the chapter on addiction, we discussed the profound effects that addictive drugs can have on the brain's reward system, including the permanent changes they make in the decision-making areas of the brain. Certain drugs can completely rewrite our priorities and reorient the brain to seek them although we know how harmful they are.

Besides recreational drugs, there have been attempts to use certain psychoactive chemicals to break down free will. From the 1950s to the 1970s, there was a covert and highly unethical Central Intelligence Agency (CIA) program in the United States called MK-Ultra. Fueled by Cold War paranoia, the goal of the program was to develop techniques for mind control and interrogation. The program explored using drugs, including LSD, as well as hypnosis, sensory deprivation, electrical stimulation, and other mind control methods.

At its peak, over eighty universities and institutions were involved in MK-Ultra experiments. Some of the experiments were done on innocent civilians, causing lasting harm. For example, LSD (lysergic acid diethylamide, a drug that can cause hallucinations and altered consciousness) was often given to participants without their knowledge. In other experiments, heroin addicts were coerced into taking mind-altering drugs, with more heroin as their reward. The program was eventually exposed and ended in 1973, but not before all the records were destroyed, making it impossible to truly know how far it went.

Besides innate behavior and drugs, another thing that challenges the idea that we are always in control of our actions is brain development.

If there were a brain region that scientists could point to as the center of free will, it would be the frontal lobes (the area right behind your forehead). The frontal lobes were among the last structures in the human brain to evolve, which means they are a big part of what sets us apart from other animals.

In humans, the frontal lobes are also some of the last structures to fully develop as we grow, reaching complete maturity only when we're well into our twenties. Before that, the neurons in this area are still making the connections they need to have a properly functioning circuit. So, for the first twenty years or so of your life, your brain isn't fully mature.*

The idea that our frontal lobes take a long time to mature is why most countries today have separate systems of justice for adolescents and adults. If you commit a crime and are under eighteen, the focus is on rehabilitation and giving you another chance. If you commit a crime and are over eighteen, you enter the adult system, which focuses more on punishment and accountability. In the U.S., this dual system was established over a century ago, back before we knew much about brain development. Still, they had the intuitive sense that children and teens were less responsible for the crimes they commit.

Most legal systems also take into consideration whether a person was truly "in control" of their actions at the time the crime was committed. These are typically called "insanity" defenses, but

* This is a good justification for making the legal drinking age twenty-one, though other factors, including military service, may explain why the legal voting age is eighteen.

251

really, they are "I wasn't in control" defenses. It is basically the defendant claiming they didn't have free will.

Another interesting situation regarding free will involves brain injury. Brain injury can occur from getting hit on the head too hard, or from a stroke, a tumor, or a disease that causes your neurons to degrade. These injuries can affect the activity of neurons in the area where they occur, or they can affect areas connected to the damaged tissue. If an injury affects your frontal lobes, it can impact your decision-making, your memory, and your ability to focus. And in some cases, it can lead to antisocial behavior.

In one study, scientists looked at seventeen patients where a known brain injury was related to criminal behavior. In fifteen of the cases, the person had no criminal record and was not believed to be antisocial before the brain injury. But after they had the injury, they started to engage in illegal activity.

In the other two cases, the injury and the criminal behavior were related because the patient *stopped* their criminal acts after the brain injury was treated and cured.

252

The study found that the most common type of injury that led to criminal behavior was a lesion on the frontal lobes. Specifically, criminality seemed to happen when a particular network of brain areas within the frontal lobes was damaged. This network of areas includes the inferior orbitofrontal cortex and the anterior temporal lobes. Scientists think this brain circuit is important for making moral decisions.

Of course, this doesn't mean that all criminals have brain injuries, or that having this type of brain injury will make you commit crimes. But it does suggest that in some cases, injuries to key brain areas can subvert your ability to make good choices.

Can Someone Hijack Our Free Will?

The next question we can ask about free will is whether it can be hacked. That is, could someone force us to do something we don't want to do?

There are several examples of this in nature, most notably the occurrence of what some people call zombies. Zombies have been a terrifying presence in horror movies since the 1960s. The concept of a dead person wandering around aimlessly, consumed by an insatiable hunger for human flesh or brains, is pretty scary.

To be clear, there are no "un-dead" zombies in real life. But if you're an insect, a kind of "living death" situation is possible. In parts of Africa and Asia, the jewel wasp wields a fearsome weapon: a venom that can control the brains of cockroaches. When the wasp injects this venom into a roach's brain, the roach is put into a trancelike state and is left completely compliant. The wasp is then able to lead the mind-controlled roach to a hole, pulling it like a horse by one of its antennae.

Once inside the hole, the wasp attaches a wasp egg to the cockroach's body, then leaves and seals the hole shut, dooming the cockroach. When the baby wasp hatches, it starts biting into the sedated roach, sucking out its fluids and then burrowing into its body to eat its insides. The larva then builds a cocoon within the cockroach, and forty days later, it bursts out of its host's carcass as an adult wasp, ready to hunt for another cockroach victim.

And you thought regular zombies were scary.

This is another example of a really complex behavior that is preprogrammed in an animal's brain. The mother wasps are gone by the time the baby wasps emerge from the cockroach carcass, yet the babies somehow know to do all of this without anyone teaching them how. Moreover, what they are doing is very intricate: the wasp has to perform brain surgery on the cockroach, inserting its stinger through the cockroach's neck and injecting the venom into a specific area of the brain.

But more important to our discussion, this is an example of one organism subverting the will of another. No one would say that the cockroach voluntarily chose to get trapped and eaten from the inside out.

This is the only time I've ever felt bad for a cockroach.

The frightening part is how this is all done with venom. The wasp's venom is keyed to specific molecules in the cockroach's brain. The venom contains molecules that mimic dopamine, which scientists think is what makes the cockroach clean itself before being taken into the hole. That's right: the cockroach grooms itself before being eaten.

The venom also has gamma-aminobutyric acid, or GABA, an amino acid that is a transmitter molecule in the brain. Scientists think the GABA in the venom suppresses brain circuits in the cockroach that would normally make it want to run away.

There are other examples of mind control in nature. In dense tropical forests, a fungus called *Ophiocordyceps unilateralis* has evolved to hijack the minds of carpenter ants.* The fungus infects the ants and commandeers their brain, compelling them to look for the ideal conditions of temperature and humidity for the fungus to grow and spread.

* You might recognize *Ophiocordyceps* as the basis for the zombies in the popular video game and TV series *The Last of Us.*

Once infected by the fungus, the ants are driven to find a plant near an ant trail and climb up. There, a fungal stalk called a fruiting body gruesomely erupts from the base of the zombified ant's head and releases spores onto the trail to infect the next cycle of ants.

In this case, the poor ants are completely taken over by the fungus's evolutionary drive to reproduce.

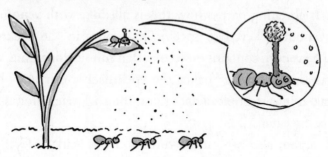

Can the same thing happen to us? In mammals, there are examples of similar zombielike brain takeovers. *Toxoplasma gondii* is a one-celled parasite that prefers to infect cats (wild and domestic cats are the only hosts in which it can reproduce sexually). One way it gets to cats is by infecting small animals like mice and rats.* When a mouse or rat is infected by *Toxoplasma,* the parasite affects its neurons, making it less fearful of predators. The parasite does this by increasing dopamine production in the host's brain, and by disrupting its fear circuits, including the amygdala. The infection even makes rodents ignore the smell of cat urine, which is typically a warning sign that a cat is nearby.

Hello, Friend!

* The parasite can reproduce inside other animals, but only asexually (by cloning itself).

256

This change in behavior makes the infected rodent more likely to be eaten by a cat, where the parasite can complete its life cycle. Humans can also be infected by this parasite, and it's possible that YOU are infected right now (especially if you have a cat and clean its litter box). The effects on humans aren't fully known, though. In some cases, infection by *Toxoplasma,* or toxoplasmosis, can cause miscarriage of developing fetuses, and it's also been linked to gray matter loss in schizophrenia patients. For the rest of us, the effects might be too subtle to detect, though if you find yourself all of a sudden liking cats more, the reason might be in your gut.

A second example in mammals is the deadly rabies virus. Rabies is transmitted through bites and saliva, and once inside the brain, it hijacks it by disrupting synaptic transmission. The result can be "furious" rabies, where the host becomes restless, agitated, and prone to biting others, with excessive saliva and foaming at the mouth. This makes it more likely for the virus to spread to others. If you've seen zombie movies, this behavior may seem very familiar.

As we mentioned, drug addiction can rewire your brain, but drugs that can control your mind, or make your mind controllable, don't really exist (the MK-Ultra experiments are generally thought to have failed). The strange thing is that someone may not need chemicals or a virus or a parasite to influence your mind.

257

In the late eighteenth century, a young man named Franz Mesmer was pursuing a medical degree in Vienna. Based on his studies, he claimed there was a kind of force within the body called "animal magnetism" that flowed like a fluid and could be harnessed to promote health and well-being. More important, he claimed that he could *control* it.

Mesmer was a charismatic figure, and he began to attract devoted followers who claimed that his techniques with magnets had cured them of a variety of ailments, from headaches and digestive issues to serious conditions like paralysis and seizures. His reputation grew, and he started to gain a kind of cult following.

In 1778, Mesmer moved to Paris, where he became a fixture in the city's social scene, entertaining guests at parties and getting a lot of attention from intellectuals and aristocrats. Despite his success, though, many in the medical establishment suspected he was a quack and a charlatan.

Enter Benjamin Franklin, the first American ambassador to France during America's Revolutionary War.

Franklin was recruited by King Louis XVI to look into Mesmer's activities. Louis figured that Mesmer had either discovered something new and amazing about the human body—or he might be a fraud.

Along with other luminary figures of the time, including Antoine de Lavoisier (responsible for much of modern chemistry and establishing the law of conservation of mass) and Joseph-Ignace Guillotin (for whom the guillotine was named), Franklin set about to test Mesmer's methods in his home outside Paris.

One of the tests involved examining the effects of magnets on a group of patients. Franklin and Lavoisier found that the magnets had no physical effect on the patients, yet some patients would still react to them. When they switched to fake magnets (basically, placebo magnets), the patients *still* reacted. In fact, they were astonished at what they could get patients to believe simply by copying Mesmer's theatricality and bogus rituals.

For example, they would blindfold subjects and pretend to apply the magnets to the subject's body parts. In this state, they could get subjects to feel pain and heat. In some cases, they could get subjects to panic and convulse, and even lose their ability to speak.

Franklin and his group credited all of this to the power of *suggestion*. Basically, they had succeeded in manipulating the imagination of their subjects.

Today, we recognize this phenomenon as *hypnotism*. Modern studies of hypnosis have found that it's possible to enter a state in which all your attention is focused inward, similar to meditation. Some scientists have found that this hypnotic state alters the way the brain communicates with itself, and that this makes you more open to the things people suggest to you.

In a recent study, scientists began by identifying people who were highly hypnotizable, as well as others who were not, and used fMRI to observe their brain activity. The researchers discovered that highly hypnotizable people under hypnosis had less activity in a part of the brain called the dorsal anterior cingulate cortex. This is an area that helps process real-time emotions and physical pain. This suggests that during hypnosis, highly hypnotizable people may be so absorbed in the experience that they are not attending to what is happening to them at the moment. Scientists also observed that some brain areas talk less to each other: an area called the dorsolateral prefrontal cortex communicated less with two other areas, the medial prefrontal cortex and the posterior cingulate cortex. This has the effect of separating your actions from your *awareness* of your actions, which might help explain how a susceptible person can go on "autopilot" when responding to hypnotic suggestions.

Fortunately, not everyone is susceptible to being hypnotized. Experts say that only about 20 percent of us respond to it, while everyone else responds to it very little or not at all. Still, the potential is there, at least for some of us, to be hypnotized and more easily guided, essentially taking away free will.

I love this book.

Another way someone can hack your brain is through direct stimulation of your neurons. Remember the scientist Delgado and his bullring experiment? Imagine what would happen if someone inserted wires into *your* brain. Could they control your actions?

Over the years, scientists have found that inserting wires deep in the brain and applying electricity to them can help in the treatment of Parkinson's disease. Parkinson's happens when neurons in your basal ganglia, a structure in the middle of your brain that makes dopamine, begin to die. With less dopamine in your system, a lot of brain areas stop working properly, especially the ones that help control your movements. Symptoms include tremors (shaking of the hands), stiffness, and seizures. But stimulating the brain with wires seems to alleviate these symptoms.

Interestingly, scientists have found that stimulating the brain can have other effects as well. In one study, scientists looked at seventeen Parkinson's patients who had wires implanted in their brain and tested how they made decisions. In the experiment, the patients were shown two symbols on a screen and were asked to pick one.

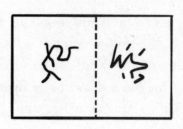

For the purposes of the study, some symbols were considered "better" than others; that is, when patients picked one, it had a higher probability of giving the patients a positive or "correct" signal. But some symbols were "worse": when picked, they had a lower probability of telling you that you were right. After a while, the patients generally learned which symbols were good and which were bad.

Then came the testing phase. The scientists looked at how long it took patients to decide between different combinations of good and bad symbols. Normally, when you have to decide between a good option and a bad option (say, picking your favorite ice cream flavor over a flavor you hate), we call that a "no-brainer." In that situation, it doesn't take us very long to decide. This was confirmed in the experiment: when asked to choose between a "good" symbol and a "bad" symbol, most patients made that decision quickly.

Duh.

GOOD

BAD

But what if you have to choose between two good options? Normally in this situation, people tend to take longer to decide. For example, let's say you were given a choice between going on an island vacation in Hawaii or going on an island vacation in Ibiza.

Which would you choose? They both sound great, right? It would be normal to take a little time to consider which one you prefer.

This was also confirmed in the experiment. When asked to choose between two good symbols, most patients took longer to make that decision.

But not always. When the scientists activated the wires implanted in the patients' brains, they found that the patients made those decisions faster. They didn't have any new information, they just acted faster and more rashly. And when they turned off the wires, the patients went back to being more deliberative.

At the push of a button, the subjects would go from being thoughtful and reserved to being more impulsive.

In this case, scientists think that the brain area where the wires are implanted, a region called the subthalamic nucleus, is important in setting a threshold for making decisions. The hypothesis is that when we are considering two options, we evaluate them and we compare their relative value. And when that difference exceeds a certain value—that is, when it's clear that one option is better than the other—we make the decision. But people who are impulsive ignore that threshold, or at least their threshold is set much lower than other people's. Basically, the scientists found that people can be "jolted" into action, literally!

There are other ways to influence your decision-making process, and some of them don't require inserting metal wires into your brain. Transcranial magnetic stimulation (TMS) is a technology that uses an electromagnet to focus and send magnetic pulses into your brain. The magnetic pulses create tiny electrical currents, which can stimulate or suppress nerve cells in the focus region. It's essentially a remote control signal scrambler that works without having to open up your skull.

TMS is being explored for treating brain disorders such as depression, drug and alcohol dependence, and schizophrenia, and it has also been used in experiments regarding how you make decisions. In one experiment, subjects were given the option to participate in two lotteries: Lottery A and Lottery B. Lottery A had low stakes, which means that if they lost, they wouldn't lose a lot of money, but it was also low reward, which means that if they won, they wouldn't win a big amount. Lottery B had higher stakes: if they won, they would get a big prize, but if they lost, the cost would be bigger.

The scientists found that when they scrambled the subjects' brains with TMS, the subjects were more likely to go with the safer bet. Scientists think that the area that was being scrambled, the dorsolateral prefrontal cortex, helps us estimate risk and reward. Once again, like Delgado's bull, the decisions we make can be influenced at the touch of a button.

Is Your Brain Predictable?

The last question we can ask about free will is whether the brain's behavior can be predicted or not. As we've seen, it's hard to say whether we are truly in control of our actions most of the time, and it's also possible for someone else to control or influence our decisions. But the real mystery of free will is whether our choices are predetermined or not.

When you make a decision, is it truly the case that you could have made a different choice? Or were you always going to make that choice? Could someone have predicted what you were going to do?

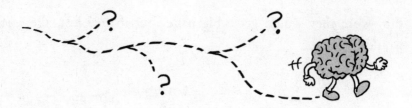

In some sense, science has shown that we *are* predictable, at least in the short term. One study looked at activity in the prefrontal cortex and whether it can predict what we're going to decide. The subjects lay down in an fMRI machine and were asked to press one of two buttons: one on the left or one on the right. They were also shown a clock and were asked to mark the time it took to decide which of the buttons they were going to press. The scientists found that activity in the prefrontal cortex and other areas predicted the choices the subjects would make up to ten seconds earlier than the time subjects reported making the decision in. In other words, scientists could tell what button the person was going to press as much as a full ten seconds before they were even *aware* they had decided.

Of course, this just means that our brains sometimes make decisions for us before we know it. In the end, you could say it's still your brain that's deciding, whether or not you are conscious of it. A deeper question to ask, then, is whether *all* your decisions can be predicted. Could someone with perfect knowledge of everything inside your head know exactly what you're going to do in the future?

This is a question that's been debated since ancient times, starting with the idea of determinism. One of the most notable Stoic philosophers (remember them from chapter 8?) from the first and second centuries was the Greek Epictetus. Epictetus believed that individuals have the ability to choose what to do, but that our choices are limited by what the world presents to us. That is, we have free will, but we live at the mercy of a world with physical laws.

Others have argued that free will doesn't exist at all. Paul-Henri Thiry, an eighteenth-century Franco-German philosopher, contended that our brains are biological in nature, and like all biology, they also follow the laws of physics. To Thiry, our brains were machines and we had as much free will as, say, a mechanical clock, or a rock rolling down a hill. In his view, humans were no different from wind-up toys, destined to behave in predictable ways.

This idea, that there is no uncertainty in the universe, is called determinism. In the 1600s, Isaac Newton discovered that everything in nature seems to follow certain laws (he called them the laws of motion) that let you predict what will happen if you throw a ball in the air, or how something will move if you push it with a certain force. Scientists extended this idea to say that if everything is predictable, then our brains are also predictable. And if our brains are predictable, then free will doesn't exist. Basically, you never have a choice. Your brain simply follows the laws of physics, just like balls or rocks do.

Today we know that this is partially true. Your brain *is* like a giant machine. As we saw in chapter 4 ("Will an AI Take My Job?"), it's made out of individual cells called neurons that are connected together like a computer circuit.

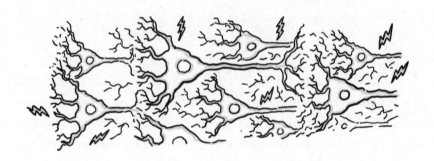

It's an extremely complex circuit—each neuron is, on average, connected to 10,000 other neurons, and each connection is different (some are stronger than others). But in the end, that's all the brain is: a giant network of simple parts.

This means that whether free will exists or not comes down to a simple question: Is a *single* neuron predictable?

If it is, then free will doesn't exist because all of your thoughts and actions would be the result of small parts acting predictably. Someone with complete knowledge of your brain could tell what you are going to do ten seconds from now, and even ten years from now. Every choice you make would be predetermined, dictated by the mechanical workings of your neurons.

Fortunately, Isaac Newton wasn't quite right. He was correct in that the universe seems to follow certain laws, but the laws that he discovered don't always work. When you go down to the level of atoms and small particles of matter, things behave very differently, and they follow the laws of quantum mechanics.

Quantum mechanics says that particles and atoms don't act like the objects we're familiar with (balls, rocks, etc.). At the smallest levels, bits of matter are . . . *fuzzier.* That is, they have an inherent uncertainty to them.

For example, you can never quite tell where an electron is, and where exactly it's going. And when you interact with it, you never quite know how it's going to react. When you try to poke an electron, there's a probability that it might veer off to the right, or it might veer off to the left. Which way it goes is totally random.

The reason you don't notice these weird quantum effects in your everyday life is that they average out for large objects. A ball or a rock has quadrillions of atoms in it, and each atom is doing something a little bit random, but on average they all move together as one thing.

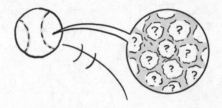

Does this mean that neurons are random? This is hard to tell. A neuron is small—a few thousandths or a few hundredths of a millimeter—but compared to atoms they are pretty huge. A typical neuron is made up of hundreds of trillions of atoms. But how a neuron reacts *can* depend on single molecules.

Remember from chapter 4 that a neuron acts like a switch, or a trigger. If the inputs it receives add up to a certain value, the neuron sends a signal to other neurons. If the inputs don't reach that value, the neuron stays silent. A neuron is like a balance scale: it leans to one side until you start to add weights to the other side. Once enough input is received, the scale tips over, activating the neuron.

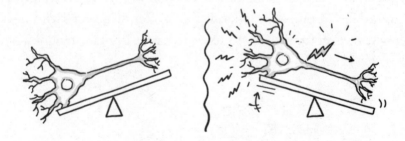

And because the output of the neuron is so binary—it either activates or doesn't—a single molecule can make the difference in what the neuron does. Signals between neurons are transmitted by molecules called neurotransmitters. These molecules are released by one neuron and picked up by chemical receptors in the receiving neuron. A single molecule might be enough to trigger the signal that is sent to the main body of the neuron.

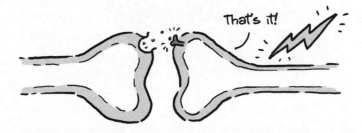

Since all this signaling happens at the molecular level, it is technically possible for quantum uncertainty to play a role. You might imagine a single molecule interacting with a single receptor. If that interaction happens to make the difference between a neuron activating or not, then the uncertainty in that molecule can make the neuron's output random. And if that neuron happens to be the deciding factor in how you make a decision in your life, that means your choice at that moment is also random.

Some scientists have proposed this as proof that we have free will. After all, if nobody can predict what we are going to decide, then the choice really lies within us.

But other scientists are more skeptical. They argue that it's highly unlikely that one molecule can determine the fate of your decisions. There are thousands of molecule receptors at the interface between two neurons, and it all happens in a messy, crowded environment. On top of that, scientists think your brain rarely depends on the actions of a single neuron. Your brain has redundancies, and scientists think it's the action of large groups of neurons that really make up your thought processes.

Unfortunately, nobody's come up with an experiment to test whether or not neurons are quantum mechanical. It might be that there is randomness in your thoughts and actions, in which case you could say that free will exists. Or it might be that your brain washes out any uncertainty, and neurons are predictable in the same way that the gears of a clock can be predicted. In that case, free will does not exist, and all of our actions are predetermined.

Making Good Choices

The jury is still out on free will. And when it comes to our brains, the concept seems to be malleable. Whether or not we are in control of our actions depends on many factors. Sometimes it can be said that we have a choice, and sometimes we can't be held responsible for our actions. It could be that we are simply following a preprogrammed behavior, or that someone else is influencing our decisions.

The weird thing is that most of us *feel* like we have free will. We all have a sense that the decisions we make are our own, and that when we make a choice, we make it in the moment.

Of course, that could just be an illusion. It could be that we are not really in control, and that free will is just a story we tell ourselves to feel better. Our brains might be biological machines and we are just robots having the *experience* of free will.

Or it could be that quantum mechanics gives our brain randomness, in which case we can say that we have free will because no one can predict what we are going to think or do.

In the end, it may not matter. Whether we have free will, or we just *think* we have free will, may not make a difference in our experience of life. Perhaps the best choice we can make is to just enjoy the ride.

Woo-hoo!

WHAT MAKES SOMETHING FUNNY?

Humor and laughter are universal to the human condition.

Humor strengthens bonds within a social group. It can even be seen in our primate cousins.

The earliest jokes date back thousands of years, often with the same themes we find funny today.

"Pull my finger"?

But what makes our brains giggle?

It's a slippery subject.

Scientists tend to separate the perception of humor (mirth) from the behavioral response (laughter).

Finding something funny is a mental state defined by delighted exhilaration.

It seems to be triggered when we experience sudden violations of our expectations.

For example, in infants, playing peek-a-boo elicits laughter.

Peek-a-boo!

This is because of the surprise of seeing something disappear.

Giggle!

But the game doesn't usually work on adults.

#1 PARENT →

They know (or should know) that objects exist even when they're not seen.

Splat!

You can also get someone to laugh by tickling them, which is a reflex protecting sensitive areas of your body.

Scientists think it's a defensive mechanism because we can't tickle ourselves.

And even though we laugh, tickling is not always pleasant.

Funnily enough, while we can't tickle ourselves, scientists have found we can be tickled by robots!

There are unfunny modes of laughter, though.

"Nervous laughter" may be a self-soothing protection against awkward social situations.

uh, heh heh...

And some people can't stop themselves from laughing inappropriately, which may be caused by an emotional disturbance called pseudobulbar affect.

HA HA HA!!!

Other people have a fear of being laughed at, called gelotophobia.

Don't laugh.

But while it's easy to make someone laugh, the perception of humor, called mirth, is much more complicated. Scientists think each kind of humor involves a different brain circuit:

TICKLING recruits regions of the cortex and hypothalamus to turn on the periaqueductal gray (PAG), which is the part of the brainstem that controls vocalizations.

"Getting" JOKES seems to depend more on language areas in the temporal lobe, the prefrontal cortex, and cerebellum.

Reading COMIC STRIPS or watching comedians turns on the frontal lobes and the reward system in the ventral striatum and the nucleus accumbens.

It's a pMRI: a pun-tional Magnetic Resonance Image.

Scientists have found that humor and laughing lowers your stress hormones.

It may also lead to the release of endorphins, which can improve your mood.

In one experiment, scientists had subjects watch comedy movies and stand-up performances.

They found that subjects had a much higher tolerance for pain afterward.

Humor may even enhance learning.

In another study, scientists took a regular psychology course and enhanced it with humor.

They added jokes, cartoons, and funny top-ten lists to each lesson.

HA!

They found that students showed more engagement and greater satisfaction with the course when humor was added.

Humor is one of the most complex and mysterious behaviors we experience.

But it definitely contributes to well-being and social bonding.

So here's the punch line: Don't jest sit there, get laughing. Humor as medicine is no joke.

Chapter 10

WHAT HAPPENS WHEN WE DIE?

The boundaries which divide Life from Death are at best shadowy and vague. Who shall say where the one ends, and where the other begins?

—Edgar Allan Poe

Why are we so afraid of dying? At first, the answer seems simple: Our evolution probably favored traits that promoted survival and reproduction, so it makes sense that self-preservation is central to our being. In other words, we evolved to not want to die.

But as brains became more complex and self-aware, death probably took on a new dimension as we understood what it means: the permanent absence of others, and of self. This is especially worrisome to us because we are, after all, pretty invested in ourselves. We work hard to build a life, form social connections, be there for our families. It can be hard to think about a world where we're not there.

It's also hard because our brains are wired to recognize how others feel and we understand

that a person's death causes distress and grief in those connected to them. So it seems natural that we feel both a biological imperative to survive and a sense of loss at the thought of our own demise.

In 1973, Ernest Becker, a cultural anthropologist, published *The Denial of Death,* which dove headfirst into our fear of dying. Knowing how crippling this fear can be, he observed how most people seem to "put a lid" on this fear and ignore it in order to function in their daily lives. Becker proposed that humans construct "symbolic selves," that is, idealized versions of ourselves, in order to protect ourselves from this fear. These symbolic selves take the form of stories that become our personal identity. For example, a parent may find comfort and self-esteem in being a parent, and they may feel a sense of purpose in the cultural narrative around the virtues of parenthood.

He also noticed that people create "legacy projects" for themselves, that is, projects that give us the comforting idea that what we do will live on after we die. If you're someone whose symbolic self is to be a parent, then your project might be to create a legacy of children who will outlive you. If you're a businessperson, you might find comfort in creating a successful business that lasts for a long time. If you're a religious person, you might invest in a devout lifestyle with the hope that doing good may provide an enduring legacy (along with the possibility of entering heaven). According to Becker, such "legacy projects" shield us from the terror of contemplating our final end.

These observations align pretty well with the ideas we explored in chapter 8 ("What Makes Us Happy?"). Finding a purpose for your life can be a strong wellspring from which you can draw contentment.

Scientists have also found that thoughts of death create strong responses in the brain's emotional circuitry, and there's evidence that the fear of death is hard-coded in your brain. In one study performed in Germany, subjects were prompted to think about death and the possibility of dying while their brains were being scanned. To prompt them, they were asked to agree or disagree with statements that came from a Fear of Death Scale (remember psychological scales from chapter 2?). A typical statement might say something like "I am afraid of a painful death," and participants would have to press a button depending on whether they agreed or not. As a control, the same experiment was done but with statements related to going to the dentist. A typical statement there might say "I panic when I sit in the dentist's chair."

"I panic when I lie inside an fMRI machine."

YES!

The study found that several brain areas that are important for processing strong emotions activated, such as the right amygdala and an area called the rostral anterior cingulate cortex. Another area, the right caudate nucleus, also responded more strongly to ideas of mortality compared to the dentist scenario. This area is interesting because it's typically associated with unconsciously working toward a goal, and also the experience of love. The scientists speculate that perhaps our brains are preprogrammed to find comfort in having a larger purpose and close relationships as a way to deal with the possibility of death.

Other studies have found that an area called the insula is active when processing the fear of death. The insula is involved in understanding how you feel about yourself. Interestingly, scientists have found that the insula's reaction to the thought of dying depends on self-esteem. In one study, scientists found that people with high self-esteem had lower activation of the insula when prompted to think about their mortality, whereas people with low self-esteem had higher activation of the insula. Basically, the more satisfied you are with yourself, the less anxiety you have about the possibility of dying.

Given that it's such a source of dread and fear, you might wonder what actually happens to your brain when you die. When does death occur, and how does it feel?

What Happens When We Die?

In 1907, a Massachusetts doctor, Duncan McDougall, performed a series of experiments to determine what happens to us when we die.

In one of the most unusual experiments of the time, McDougall devised a scale large enough to support a patient. The patients he chose for his experiments were near death and McDougall hoped to weigh them before and just after the exact moment of death so that he could capture the change in weight when the person died.

McDougall reasoned that the cessation of brain function and physical death should trigger a change that could be measured. His thinking was based on Descartes's idea that the body and the mind were separate things. Therefore, he argued, when we die our "soul" must go somewhere, leaving the physical body empty.

When he did the experiment, McDougall and his team observed a slight drop in weight that they could not attribute to evaporation or other loss of body fluids. The difference was 0.75 ounce, or 21 grams, which McDougall proposed was the exact weight of our spirit, or "soul substance."

Today, we know that McDougall's experiment was terribly flawed. It was founded on faulty assumptions about the nature of brain function, it was full of experimenter bias, and it was based on a total of only four patients. Still, the durability of the 21 grams story reflects our profound curiosity about what happens to us when we die.

McDougall went on to attempt other strange pseudoscientific experiments—including attempting to photograph souls leaving the body at the moment of death. His efforts reflect a search for comfort in the face of the cold fact of our own demise, and a way to quantify what is lost when we "leave" the material world.

What happens to our brain when we die is still a source of debate, though. Today, most scientists don't think that anything supernatural occurs, but there is still uncertainty about how your brain reacts in your final moments.

Part of the uncertainty is that it's hard to define when you actually die. Humans haven't always been 100 percent reliable in determining when someone is truly and completely deceased.

One notorious case happened in 1650 involving a woman named Anne Greene. On December 14 of that year, Anne Greene was brought to the gallows in Oxford, England, where she was hung for the crime of infanticide (in reality, she had just miscarried). The witness account states that she was hung by being separated from a ladder, which chokes you. This is different from being dropped from a retractable platform, which typically breaks your neck. This grim detail might explain what happened next.

At the time, the anatomy professors at Oxford University would collect the bodies of executed criminals as teaching aids for dissection in the medical school. When they opened Greene's coffin for dissection the following day, they were met with a stunning surprise—she was breathing, and had a faint pulse. Apparently, Greene had been strangled unconscious by the rope, but her brain received enough oxygen to keep her barely, imperceptibly, alive. Greene was treated, recovered, and was ultimately pardoned because, it was reasoned, the hand of God must have spared her. She later married, had three children, and lived another nine years.

Before you think this kind of thing happened only in the distant past, consider that cases of people being declared dead only to be found later to be alive still happen, even today. Usually, it's the result of medical errors, such as failing to administer a thorough neurological examination or misinterpreting its results.

Determining death is a very important problem that physicians encounter on a daily basis. What you might call "brain death" can be very different from what you might call "clinical death" (which is when your heart stops beating), and both of these can be different from biological death (which is when your cells start dying) or even legal death (which is when there is no hope of resuscitation).

This uncertainty is especially true in cases where the patient is connected to machines that can preserve the function of the rest of the body even after the brain is dead, sometimes for many months.

So what actually happens when our brains die? Well, the brain uses about 20 percent of the body's oxygen, and without it, it starves. Interrupting the flow of oxygen-bearing blood to the brain for even a few minutes (for example, because of an accident or a stroke) can result in the death of your neurons. The lack of oxygen causes your cells to lose their ability to control ions coming in and out of them, throwing everything out of balance. These changes cause calcium to accumulate inside the cell, which then triggers proteins and free radicals that start to shut the cell down and break everything apart. And all of this can happen within minutes.

There are exceptions, though. In 1988, a two-year-old girl, Michelle Funk, fell into an icy creek near Salt Lake City and was submerged for over an hour before she was brought to the surface, seemingly dead. Miraculously, Michelle Funk lived. Rescuers immediately administered CPR after recovering her from the water, and she was brought to a hospital where doctors used a heart-lung machine to warm her blood outside her body.

Doctors think Michelle was saved by hypothermia, a lowering of core body temperature below 35°C (95°F). Icing your body essentially stops cell metabolism and reduces brain swelling and inflammation, which can prevent cell death.

A sudden lowering of your temperature may also elicit a reflex in your body called the "diving response." It's an evolutionarily old reflex seen in aquatic mammals that lowers the heart rate and constricts blood vessels in order to preserve oxygen when you submerge in water. If you've ever dipped in a tub or a pool and felt a sense of calm underwater, that's the diving response in action.

Through the combination of initial hypothermia, the resiliency of her young brain, and the actions of her rescuers, Michelle survived the ordeal and went on to live a normal life. What's fascinating is that people like Michelle have experienced death, in a way, and have lived to tell about it.

Near-Death Experiences

What is it like to experience death?

You might think that death is, by definition, the end of being able to experience anything. But, as we discussed earlier, the definition of death is somewhat fuzzy. For example, Michelle's heart stopped for over an hour, and despite the beneficial effects of hypothermia in this instance, it is likely that the lack of oxygen caused damage to her brain cells. Scientists credit the fact that she was young with her brain's ability to recover or compensate for any loss of function.

It's hard to know what Michelle experienced (she was two years old at the time), but some people who have survived similar accidents have been able to report on what it felt like to be so close to death. In many of those cases, victims describe having what are called near-death experiences (NDEs) in the period during which their brain function was impaired. NDEs can be intense and meaningful, and it turns out that there are many similarities in what people say about them.

Most near-death experiences play out like this: There's a feeling of being completely free of pain and seeing a bright, glowing light at the end of a tunnel. You might feel like you're leaving your body behind, watching it from a distance or as you float up into space. You might meet a loved one who may be living or dead, and you might also encounter spiritual beings like angels. Some people report having an intense feeling of their life flashing before their eyes. Time and space get distorted, making the experience all the more surreal.

Whatever near-death experiences are, they're different from our typical conscious experience of the world.

Near-death experiences also seem to be dependent on cultural norms. Studies comparing them between Western and non-Western cultures have found some common elements, such as the impression of a bright light, but others, like seeing one's life flash before one's eyes, were not universal. This suggests that cultural expectations, or context, may lead you to experience what you expect to happen.

Is there a neurological basis for near-death experiences? It turns out that the out-of-body feeling that people feel in near-death experiences has a lot in common with what scientists call *autoscopy*, which is the feeling of observing yourself from outside your body. This can happen to people at random times, not just when they are close to dying.

In one study of five patients, the feeling of being outside their body was tied to disruption of a brain area called the temporoparietal junction, which is just above and behind your ears. In the study, scientists used transcranial magnetic stimulation (TMS) to scramble this area using focused electromagnetic pulses. When they did this, the patients had a harder time estimating where their left hand was, which was blocked from their view. Scientists think this area is important for processing information from all of your senses to give you an idea of where your body is. When this area is disrupted or not functioning well, you feel separate from your body, in the same way you do with an out-of-body experience.

So what people perceive as leaving their body near death may just be a disruption to this area of the brain. But what is causing the disruption?

Scientists think that when your brain is dying, a series of neurochemical changes are triggered. Some of these changes can account for what certain people perceive in near-death experiences.

For example, remember the brain chemical serotonin? For many years, it was thought that lower serotonin levels were responsible for depression, a view that we now know to be wrong, or at best incomplete. Yet somewhat paradoxically, serotonin seems to spike in the brain at the time of death. Scientists have measured this in small animals and found that rodents experience a threefold increase in serotonin in their brain as they die.

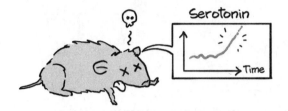

Serotonin could be an especially important clue in the sensory distortions that people feel in near-death experiences. Some drugs that cause hallucinations achieve their effects by disrupting serotonin receptors. Drugs like LSD and "magic mushrooms" (which contain the neuroactive chemical psilocybin) are thought to mainly activate serotonin 2A (5-HT2A) receptors. This boosts serotonin levels, leading to changes in perception, mood, and thought. We can only speculate, but it's possible that a spike of serotonin at death could mimic the serotonin-based effects of hallucinogens, leading to similar "trippy" or weird experiences.

Another hallucinogenic substance, N,N-Dimethyltryptamine, or DMT, might also provide a clue. DMT is the psychoactive ingredient in ayahuasca, a plant-based psychedelic used in South American shamanic traditions. DMT is structurally similar to serotonin and also binds to the same chemical receptors. Similar to serotonin, levels of DMT have been observed to spike in the visual cortex of rodents at death.

In fact, you can actually have something similar to a near-death experience if you take DMT. A study in 2018 gave healthy volunteers either a placebo or DMT, and then they were asked to describe their experience. The experiment would ask subjects if they experienced feelings of joy, or the feeling that time had stopped, or if they had the sense that they were outside their body. The scientists found that DMT significantly increased the experiences associated with near death compared to a placebo. These findings suggest that there is a striking overlap between real near-death experiences and the psychedelic state induced by DMT.

Of course, near-death experiences are unlikely to be solely explained by a handful of neurochemicals. As cells start to die, a large and complex series of metabolic changes occurs. You might expect that the brain simply switches off, but recent studies have found that the brain actually increases its activity in the moments leading up to death.

In one study, scientists recorded the brain signals of rats while their hearts stopped. They analyzed changes in different features of the signal: how strong the signals were and how synchronized the activity was across brain areas. They found that within the first thirty seconds after the heart stops, the brain surged in activity at a particular wavelength. This wavelength, called gamma oscillations, is usually a signal of conscious mental processing. These oscillations showed up across the brain, especially in connections

298

from the front to the back. The activity also seemed to be coupled to other types of brain waves called theta and alpha, which signal information processing and memory.

Then, after the thirty seconds were over, the brain flatlined and stopped working altogether. This meant that when the heart stopped, and before the brain ceased to think, a period of intense mental processing happened.

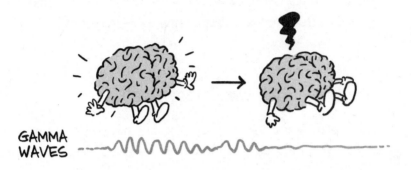

GAMMA
WAVES

Studies like these are nearly impossible to do in people. But in 2022 there was one case where a patient's brain was being analyzed and the patient unfortunately died during the recording. The eighty-seven-year-old man had arrived at the hospital after a fall and had quickly deteriorated due to a series of brain clots. The doctors had been monitoring his brain activity to look for signs of seizures when his breathing and heart rate began to fail. After discussing it with the man's family, and following his "Do not resuscitate" wishes, the doctors stopped treating him and he passed away.

Analysis of the brain recordings showed something amazing. As with the rats, the patient's brain also saw a rise in activity after the heart stopped. And again, as with the rats, the burst was in the gamma wavelength. While we can't know what the patient was thinking when he passed away, this complex, death-related activity pattern suggests that the brain is doing more than just simply switching off at the time of death.

Scientists think near-death experiences might be the brain trying to make sense of what is happening to it—a sort of "last gasp" of your dying neurons.

Can We Cheat Death?

If contemplating the inevitability of death has you thinking that you'd like to postpone it as much as possible, then you're not alone. It's natural to want a life that's long and active, and you might say that most of humanity's efforts over the last few thousand years have all been focused toward that goal. Medical science has achieved remarkable feats, and the expansion of our lifespan beyond the theoretical limit of around 125 years seems possible. Promising new avenues of research and discovery are emerging, and the future holds exciting prospects for extending the boundaries of our mortal existence.

One such avenue is gene editing. Since aging is in our genes, gene-editing tools like clustered regularly interspaced short palindromic repeats (CRISPR) and other gene therapies are quickly improving our ability to edit all sorts of genes, including those responsible for aging. It's easy to imagine that a gene-editing strategy that suppresses the genes that lead to death and disability is in our near future. These strategies might also promote, or express more of, those genes that repair cell damage, giving us a greatly extended and healthier life.

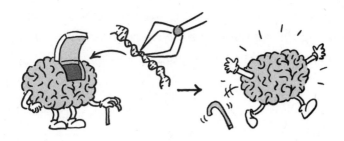

Some of the current gains in longevity or in treating age-related conditions are impressive. Experiments in which gene therapies are delivered to mice via designer viruses have been able to extend the lifespans of those mice by as much as 41 percent without negative side effects, such as promoting the growth of cancers. For us humans, that would be like adding thirty years to our lives.

While these treatments aren't being used in people yet, there have been some reports of successfully reducing aging in cells grown from people. In one of these studies, cells donated by a 114-year-old "supercentenarian" were de-aged to those of a young person using a cocktail of special genes. These genes, called Yamanaka factors, have the ability to take an old human cell and revert it back to a stem cell. Stem cells are important because they are the type of cell that embryos are initially made up of, and they have the capacity to not only become any kind of cell in your body but to regenerate cells and tissue damaged by injury or disease. Basically, we may be able to reverse-age any cell in your body back to a younger state.

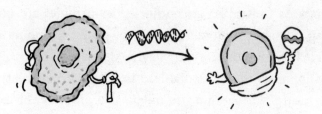

Other advances may enable us to manipulate aging in both directions, essentially rewinding or fast-forwarding your slow march to death. By manipulating what's known as the epigenome (controlling the expression of DNA), scientists have been able to use Yamanaka factors to both speed up and slow down aging.

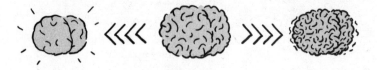

Scientists have also focused on senescent cells, which are old, stalled cells in your brain and your body that refuse to go away. These cells emit damaging molecules called inflammatory cytokines that cause inflammation and contribute to aging. Targeting these cells and finding ways to remove them in mice improves their lifespan and health. Plans for clinical trials to translate these treatments to humans are underway, though they will take a few years to design and run.

Can We Cheat (All) Death?

While science may eventually be able to delay the aging process and natural death, it may not be able to protect us from other threats like new diseases or sudden accidents. Recently, scientists have started asking if there's a different way for your brain to cheat death, one that is rooted in a common science fiction trope: Can you upload your mind into a computer?

Think about it: you might be able to achieve immortality by copying your consciousness into another form, perhaps in the body of an android or a clone. But while there have been significant advancements in neuroscience, computing power, and data storage, the brain as we've seen is an intricate and complex organ, and it's not clear that we could ever copy it completely.

For starters, what would you copy the brain into? A computer that can support the operations of a human brain and simulate its functions would need to be engineered.
To do that, several companies have been developing "neuromorphic chips" that can do computations that mimic those of the brain. Current neuromorphic chips have millions of artificial neurons, and they are the most promising route toward a neuro-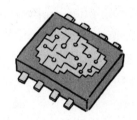
computing platform that is modular, expandable, and capable of simulating the rules of plasticity found in our own brains.

Another possibility is to clone your brain. Scientists have been successful in growing "mini brains" on a petri dish. These clusters of cells are called "cerebral organoids," and they're made from stem cells taken from people or mice. Scientists have been able to grow brain cells, record their electrical activity, and provide feedback on their performance. By doing this, scientists have been able to teach networks of cells to perform complex actions, like playing video games.

It's not unreasonable to think about this technology being able to copy the functions of an entire brain in the future.

But the real barrier to copying your brain might be in scanning it. Current brain-imaging technology like fMRI and PET scans can provide estimates of your brain's functional connections, with a resolution down to a few cubic millimeters of brain volume. That sounds impressive, but even a volume of brain tissue that small can hold over 150,000 neurons, each of which is potentially making thousands of connections with other brain cells. So it seems impossible to capture the fine-grained details of what makes you You with current imaging methods. Anatomical methods like electron microscopy or cryonics can improve this resolution down to a single synapse, but these methods require that you be dead, infused with preservatives or frozen, and your brain sliced up into thin sheets, which is . . . not fun.

And there's no evidence that even detailed anatomical scans can capture the status of the many thousands of types of proteins in each of your trillions of synapses. What can be captured with current technology is just the tip of the proverbial iceberg in terms of the massive amount of activity that determines how your brain processes information.

And then there are the philosophical questions. Think about it: If someone was able to make a perfect copy of you, would it still be "you"?

In your day-to-day life, you have the sense that you are the same consciousness. When you wake up in the morning after a night's sleep, you still think you're the same person that went to bed the night before. Coma patients that reawaken, even after years, still retain their sense of self (though some are shocked when they discover they have aged). Henry Molaison (H.M.), the famous memory patient discussed in chapter 5, had an impaired memory, but he still seemed to retain a sense of who he was in the present. Would you still have this *continuity of consciousness* if someone copied you?

Who are   you?

The problem can be better understood by imagining that instead of dying, you were able to transfer your biological consciousness to a computer while you were still alive. Now there are two copies of you. Your old copy would keep generating your consciousness and you'd still have the feeling that you are still you, in your old body. But your new copy would also feel like they are you, albeit in a new body. Which one would be the "real" you? And if your biological body died, would the new copy feel like you, or would they regard the old you as a different person?

Alas, I knew him well.

Is Death Necessary?

Science is a process of exploration that tries to expand the limits of human knowledge. Occasionally, we hit some hard limits (for example, the speed of light), but for the most part, we seem to be able to overcome the majority of obstacles that get in our way. Right now, it seems clear that we are getting closer to realizing solutions to the problem of dying.

Clearly, we have a lot to be thankful for. Thanks to science and public health, most of us in this generation will lead healthy, long lives. If you had been born around the year 1900, your average lifespan would have been in your forties, instead of the mid-seventies as it is today. These longer lives are due in part to vaccines and other public health measures that reduced child mortality, as well as treatments for diseases that would have greatly shortened our lives a few decades ago.

Nobody seems to mind living longer and healthier, and we can find purpose and meaning in our longer lives. A recent survey found that those who live past one hundred are generally happier and more content than people in their sixties.

Oh, you youngsters!

But if we somehow managed to conquer death completely, *should* we? What would happen if we succeed at cheating death, and we all significantly extended our lives, maybe to the point of never dying? What kind of impact would that have on us as a species and on the planet?

For instance, would living longer lead to a depletion of the world's resources? Would we benefit from the collective wisdom of an extended life, or would having much-older generations alive cause society to stagnate from old ideas and entrenched leadership? And, what if cheating death was so expensive that only the rich could afford it—would it create classes of genetically enhanced "lives" and "live nots"?

More profoundly, would it change us as a species? Until now, evolution has used genes for the adaptations that made us who we are. These adaptations were driven by competition in nature and natural selection. But if genes are just a code, and we can alter that code, then we are not letting nature take its course. One danger might be losing the ability to evolve new types of intelligence that may be necessary to deal with threats we can't anticipate. Would engineering our genes and our brains lead to a homogenized way of thinking?

Evolutionary forces created the brain and mind. Maybe the mind will be responsible for the next evolution of our intelligence and longevity. If that happens, let's hope we remember that the true meaning of living is not necessarily to live longer.

The End Is Here

To summarize, science is slowly figuring out how the brain functions as it dies. If science eventually yields a complete explanation of near-death experiences, there is still the question of what happens *after* our last neuron winks out. The good news is that understanding how the brain dies may provide amazing new insights into how it lives.

Death can be scary, but as with other fears and phobias, we can find some relief by discovering purpose and meaning in the time that we do have here on earth. A limited life can inspire us to live in the moment and not in fear of the future—to live a life of kindness and connection. While there's no evidence to believe our consciousness survives the death of our physical brain, having positive social relationships and doing work that has a lasting impact are ways that we can feel that who we are lives on.

Chapter 11

WHAT MAKES US HUMAN?

Weaseling out of things is important to learn. It's what separates us from the animals. Except the weasel.

—Homer Simpson

If you are reading this, then it's safe to assume that you are a human. Obviously, there is the possibility that you might be an Artificial Intelligence decoding this, or a visiting alien, but assuming you are a biological entity from this planet, we're pretty sure you're not an ant. Or a cat, or a dolphin. What you're doing right now, scanning these letters and processing the words and meaning of these sentences, can be done by only one species on Earth: us.

Humans have done pretty well for themselves. We're the only species that has managed to leave this planet,* the only ones that have invented machines to do things for us, and the only ones that have come far in understanding what the rest of the universe looks like and how it works. We're also the only species to have a cartoonist and a neuroscientist write a book about the brain with doodles and comics in it, but of course the merits of that are debatable.

How do you explain everything that humans have been able to achieve above other species? As we've seen in this book, humans are not alone in doing many of the things we usually think of as being uniquely human. Others animals love, hate, feel fear, get addicted, laugh, and can even be said to be happy. But no other species seems to have the drive and the ability to modify their surroundings on a global scale (for better or for worse), or to adapt so quickly to different climates and environments. And it's all thanks to that three-pound organ inside your head.

* At least on purpose. There is a theory that bacteria first came to Earth from Mars by hitching a ride on a meteor.

It's easy to be impressed by the human brain. It's famously said to be one of the most complex organizations of matter in the known universe. But while it has been studied a lot, in many ways we are still at the beginning of our understanding of how this squishy, wrinkled receptacle of thoughts does what it does.

What exactly makes it stand out from all the other brains in the world? That is still somewhat of a mystery, but a few obvious possibilities come to mind. Size is a clear contender. Humans have a bigger brain than most species on Earth. Complexity is another possible reason. Humans have some of the most densely packed brains of all animals. Consciousness (as we learned about in chapter 7) is another. Humans also have special brain areas that allow us to learn specific skills like language and using tools. Could these be what set humans apart and put us on a fast track to technological and societal advancement? Let's explore each of these traits and see whether they can explain humans' success on this planet.

Size Matters

Around 2.5 million years ago, our heads exploded. In size, that is. For billions of years, the brains of our ancestors had been on a steady trajectory of growth, evolving from simple clusters of nerve cells to more complex nervous systems capable of navigating new environments.

It all started with simple organisms about 3.5 billion years ago. Bacteria, the earliest known single-celled life-forms, needed a way to control how much water was inside them. Take in too much water, and your cell explodes. Take in too little, and you'd shrivel up and die. To control this, bacteria evolved a clever solution: ion channels and ion pumps. Basically, these are little protein machines on cell walls that regulate how many ions like sodium go into and out of a cell. A cell can get a lot of water to come into it by absorbing a lot of sodium because water tends to go from places where there isn't a high concentration of ions to places where there is. And the cell can let a lot of water out by expelling some of that sodium.

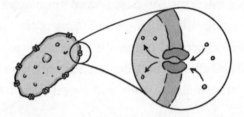

But what was initially a water management system soon became useful for something else: communication. Ions like sodium also have an electrical charge. So by moving this charge into and out of the cell, you are also creating an electrical signal. At first, bacteria used this only to send signals across its body, like, for example, to activate their little tentacle-like flagella in sequence to push themselves in the water.

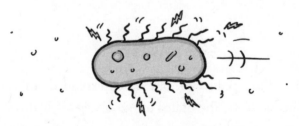

Eventually, though, simple organisms started to join and work together in colonies. About a billion years ago, tiny aquatic creatures known as choanoflagellates came about. Choanoflagellates are the closest living relatives of animals that are made up of only one cell. They have a whiplike flagellum that pushes the choanoflagellate through the water like a fish tail. Most choanoflagellate species survived quite well on their own, moving around with their tail and using it to pull bacteria in to eat.

But some choanoflagellates also formed colonies by splitting into multiple cells that stayed connected to each other, and that's when they started to communicate. These simple groupings started to evolve some of the same genes that our modern brains use for chemical signaling: genes that build proteins that concentrate and release neurotransmitters, which are essential chemicals for communication between brain cells.

These colonies likely evolved into sea sponges, which appear in the fossil record as far back as 890 million years ago. Sea sponges were some of the first animals that could stay in one place—they no longer needed to move around to hunt for food. By coordinating the actions of lots of different cells, each with their little whiplike appendages, they could pull the food to them by grabbing what they needed out of the water around them.

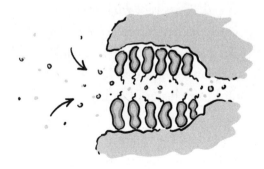

In 2010, scientists read the genome of a sponge called *Amphimedon queenslandica*, which is native to the Great Barrier Reef in Australia. It turns out that humans share as much as 70 percent of our genome with sponges. This is how important simple cell function and communication are: as much as 70 percent of our genes are devoted just to that.

Sponges represented a transition to lots of cells acting together as one. About 540 million years ago, this new way of doing things became dramatically popular in what is called the Cambrian explosion. In just 20 million years, a host of new multicellular life-forms appeared, and that's when we start to see the first signs of brains evolving.

It started with eyes. During the Cambrian explosion, organisms took the light-sensing proteins of simple creatures and started organizing them into structures that could see things accurately at a distance. The development of eyes allowed predators to spot their prey more efficiently from afar and hunt it. This probably put pressure on all organisms to be faster and more coordinated, both for prey to swim away and for predators to be smarter.

Soon, we start to see organisms with the first nervous systems and a distinctive "head." Fossils of what is considered the very first fish, called *Haikouichthys,* date from the Cambrian period. This was a smallish, 2.5-centimeter-long, jawless fish that had a vertebral column, a brain-like cluster of cells that may have been capable of "thinking," eyes, and rudimentary hearing and smelling organs. It's extinct, so we can't study its genome, or even know for sure what it could do, but it's in these kinds of bony fishes that we start to see a hallmark of all vertebrate brains like ours: myelination.

Myelin is a fatty substance that wraps and insulates nerve cells, allowing them to communicate faster and over longer distances. With this, early brains were able to grow in size without sacrificing their reaction times.

After simple fishes developed, and with the ability to grow their brains, our ancestors started to take on new challenges, which put them in situations that needed bigger and bigger thinking organs. About 360 million years ago, creatures began to leave the water and move on land in search of food or better living conditions.

Our amphibian ancestors were likely lungfish, which as the name suggests had air-breathing lungs. Senses that were adapted for an aquatic environment were tweaked to function in open air. And thermoregulation also became more important, since the capacity of air to hold heat is much less than that of water, which meant that our ancestors faced bigger variations in ambient temperature. On top of that, water, which was plentiful in aquatic environments, could be scarce on land, and being caught in a drought was a real possibility. All these challenges may have required more brain cells to process and respond to what was going on, leading to bigger and bigger brains.

Fast-forward a few hundred million years to the present, and you can ask, "How big are brains now?" Scientists have wondered about this, and they've measured the brain sizes of many different animals. For example, a finch has a brain about the size of a small pea (500 mm³), whereas a sperm whale has a brain about the size of a basketball (8,000,000 mm³). Things get particularly interesting, though, when you plot the brain size of different animals against the size of their bodies. You get a figure that looks like this:

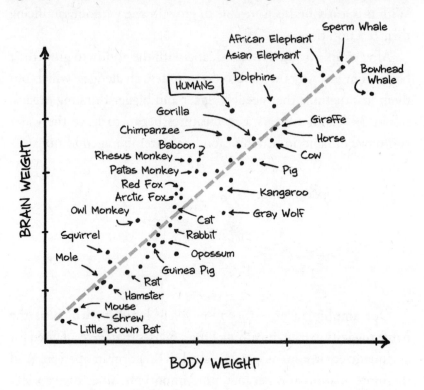

The first thing you'll notice from this graph is that humans do not have the biggest brains in nature. That honor goes to whales and elephants. Even bottlenose dolphins have a bigger brain than humans. So, while we have a larger brain than most animals, ours is still up to ten times smaller than the biggest brains out there.

Whales, dolphins, and elephants are very intelligent, maybe more so than we have realized to date. But it's clear that none of these animals can design complex computers, or figure out how to do nuclear fission, or send spacecraft to Mars, as humans have. This means that sheer size alone doesn't explain the advantage human brains have in being able to do these things.

316

The second thing you might have noticed is that all the animals in the graph fall roughly along a straight line going from the lower-left corner of the graph to the upper right. In other words, you don't see a lot of animals that have small bodies and huge brains, and you also don't see animals that have huge bodies and tiny brains. Animals seem to generally follow a simple rule: the bigger they are, the bigger their brain is.

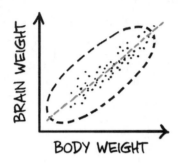

Scientists think this is mostly due to two reasons. The first is the fact that larger bodies are more complicated. That is, they have more parts like muscles and nerves to monitor and control. The second reason has to do with energy. It takes a lot of calories to keep a brain going, which means it doesn't make sense for a small animal to have a humongous brain. If it did, it would expend almost all of its energy on its thinking organ.

Because absolute brain size doesn't tell you much about how smart an animal is, scientists look at a different measure of brain size: its brain size relative to its body size. For example, does a wolf have a brain that is bigger or smaller than their body size would dictate? This is calculated most simply as the ratio of brain size to body size (brain size divided by body size).*

Scientists think that having a high brain-to-body ratio means that you have more brain matter than what you actually need to keep your body running. This "extra" brain matter presumably goes to making you smarter.

And for the most part that is the case. If you look at the graph, which mostly focuses on mammals, you'll see that primates (monkeys, apes, and humans) are all near the upper border of the cluster of data points. Primates are, on average, more intelligent than other mammals.

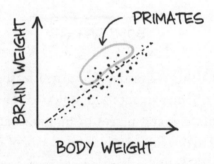

Their position near the top of the data cluster means that primates, both big and small, have larger brains than other animals that are their same size. The only exception is the gorilla, which, due to its large body size, sits right in the middle of the cluster.

Out of all the primates, humans are at the very top with the biggest brain. So, one way to explain why humans are so advanced is that we belong to a group of animals (primates) that have the biggest ratio of brain size to body size, and within that group, we have the largest brain of them all.

* There are more complicated measures, like the *encephalization quotient*, which takes into account the nonlinear relationship between brain and body sizes across species of the same group.

Having the largest brain of all the primates came about rather suddenly: 2.5 million years ago, the head "explosion" we mentioned before took place. At around that time, the brains of our direct ancestors were about the same size as a chimpanzee's brain is today. But in just a few million years, our brain capacity nearly *tripled* in size.

Here is a graph that shows the cranial capacity (as measured by skull size) of our recent evolutionary ancestors:

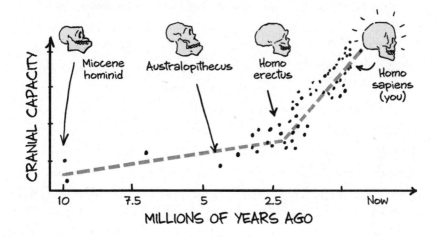

Scientists aren't sure what drove this rapid expansion of brains. One theory is that we started to walk upright with two feet. This development, as seen in the features of our ancestors' pelvic bones, happened at around the same time. It could be that walking upright opened up new opportunities that needed more brain power. For example, walking on two feet frees your arms to carry things and use them while walking. Developing these skills may have put more evolutionary pressure on our brains to grow.

Another theory is that the human primate population increased at around this time, leading to more competition and accelerated brain evolution. Two and half million years ago, scientists estimate there were fewer than 20,000 hominids (humans and our recent ancestors) on earth. Nowadays, there are almost 8 billion humans.

As the population began to grow suddenly, the increased competition, tribal conflicts, and different cultures may have put more pressure on our brains to evolve and get bigger in size.

But brain size can't be the only thing that matters. Case in point: Neanderthals had a bigger brain than modern humans. Neanderthals were a kind of hominid that evolved at the same time as our main human ancestors did. Scientists think that Neanderthals competed with prehumans, but eventually lost and went extinct. In this instance, having a larger brain didn't help. They didn't disappear completely, though. Recent genetic analysis shows that some of us today have as much as 6 percent of Neanderthal DNA.

While brain size and brain-to-body ratio tell us a lot about why humans are the dominant species on Earth, they don't give us the full picture. For instance, whales still have much larger brains than humans, and other animals like some monkeys and birds have much higher brain-to-body ratios than we do (monkeys have ratios of about 4.8 percent, while finches have ratios of 4.2 percent; humans have a ratio of 2.5 percent).

What else could account for our cranial conquest of the planet? The next likely suspect is how convoluted the brain is.

It's Complicated

As most of us learn in our lives, size isn't everything. It's the substance of things that really matters. But is that the case when it comes to *brain* matter? The answer is yes and no. One question you can ask is whether humans have the most intricately constructed brains in nature. A way to measure that is by looking at the degree of *gyration* in the brain. Simply put, this refers to how wrinkled the brain is.

When we think of brains, we usually imagine a creased and bumpy organ, but actually most brains in nature are smooth. Fishes, reptiles, and birds all have smooth-looking brains, and even mice have fairly wrinkle-free brain surfaces.

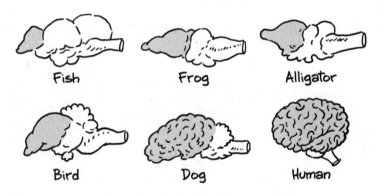

Fish   Frog   Alligator

Bird   Dog   Human

The wrinkling of brains started to happen around the time that mammals started to get bigger. Generally speaking, the larger the mammal, the more wrinkled its brain is. It's a rule that seems to cut across species. For example, primates for the most part have wrinkled brains, but small monkeys like the marmoset have smooth brains. And mice and rats have smooth brains, but large rodents like the capybara have wrinkled ones. Interestingly, scientists believe the increase in the size of mammals (and the ensuing wrinkling of brains) happened only because of Chicxulub, the asteroid that crashed on Earth 66 million years ago. Chicxulub's impact led to the extinction of dinosaurs, which cleared the way for mammals to increase in size and take over.*

See ya.

The wrinkling of the brain allows for shorter connections between its different regions, since having folds brings all the parts closer together. It's also a good way to cram more brain surface area into the same space in your skull. For example, if you "unfolded" the outer layer of the human brain, it would spread out to about the size of a 20-by-20-inch table napkin.

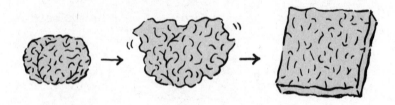

Surface area is important because that's where the brain's cortex is. The cortex is where most of your "thinking" happens, and it's made up of layers of neurons connected to each other.

---

* Another fascinating fact: scientists speculate that one of the reasons mammals survived the asteroid impact and the resulting climate catastrophe was potentially their brain's ability to sleep and hibernate. While dinosaurs had nowhere to hide, our mammalian ancestors might have been safely tucked away in underground burrows, sleeping it off until the worst of it passed.

The hypothesis is that when brains needed more neurons, the cortex expanded, and once it ran out of room it started to fold in on itself, creating the wrinkles and grooves you see in most large brains. So, do humans have the most wrinkled brains in nature? Not quite.

Looking at the neuroanatomy of whales and dolphins, it's clear that their brains are bigger than humans' AND that they are also more convoluted or wrinkled. This is what the cross section of each brain looks like:

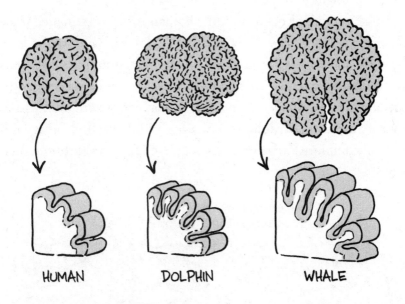

HUMAN   DOLPHIN   WHALE

If you computed the surface area of a dolphin's brain and "unfolded" it on a flat surface, it would spread to about the size of a 24-inch-by-24-inch printed newspaper (3,745 cm^2). Generally speaking, dolphin brains have a much larger surface area per volume than human brains by a factor of about 50 percent. And estimates of whale brains put their cortex surface area even higher. One study estimates that minke whale brains have a surface area of up to 6,563 cm^2, which is almost three times the area of the human cortex.

UNFOLDED BRAINS:

Humans   Dolphins   Whales

Another way to measure the complexity of a brain is to count how many neurons it has. One brain might be smaller and less wrinkled, but if it packs more neurons in the same amount of space, it might have more processing power.

The number of neurons in the human cortex is estimated to be about 16 billion. How many neurons do dolphins and whales have in their cortices? It varies. An estimate for the minke whale puts their number at 12.8 billion, which is less than in humans. An estimate of dolphins puts their number much higher, with pilot whales (which are actually dolphins) having 37 billion, and orcas (also actually dolphins) having an astounding 43 billion neurons in their cortex.

Remember when we made our first billion?

Clearly, the sheer number of neurons can't explain why humans appear to be more advanced. Could the difference be in how the neurons are connected to each other? One clue is that whales and dolphins don't have as many layers of connections as humans do. Human brain cortex is arranged in six layers of neurons (aptly named layers I, II, III, IV, V, and VI). Whales and dolphins, however, have only five layers (they are missing layer IV). This gives human brains an enormous advantage, regardless of size, wrinkles, or number of neurons. Having an extra layer might be like having a whole extra computation step, allowing the brain to do more complex calculations.

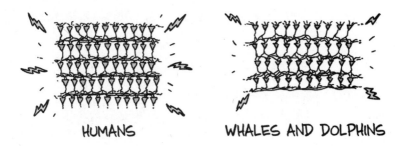

HUMANS   WHALES AND DOLPHINS

In the end, scientists think that it's a combination of size, wrinkles, neuron count, *and* connection layers that give humans the edge over other species. Some scientists have proposed a measure that considers all of those factors, called information processing capacity (IPC). IPC takes into account the number of neurons in the cortex, how close together they are, and how quickly they can send signals to each other. By this rough standard, you get a ranking that closely follows what we think of as intelligence. For example, humans rank highest in IPC because they have a good combination of all these traits. But whales and dolphins, for instance, are lacking in some of these areas (they have sparser brains and slower neurons), despite having a larger size and, in some cases, more neurons.

A Skilled Brain

Finding a measure like IPC that ranks humans highest tells you what brain characteristics seem to be important in deciding intelligence. But it doesn't tell you WHY our brains have those characteristics. In particular, it doesn't tell you what gave us the edge over our prehuman relatives. Apes and other primates also rank high in IPC: they have similar numbers of neurons, connectivity, and speed. Therefore, a final question we can ask is "What made humans stand out from the rest of the primates?"

One hint is a gene called ARHGAP11B. It's a gene that's found only in *Homo sapiens* (us), Neanderthals, and Denisovians, another one of our recent hominid relatives. ARHGAP11B seems to act on the metabolism of neurons in the cortex, causing them to grow and multiply. Normally, the way scientists would confirm this is by deleting the gene in a test animal and seeing what happens. But in this case, it's hard to do that without causing side effects that might confuse what the gene actually does. So scientists decided to test the hypothesis by putting the gene into the brain of a species that *didn't* normally have it: the marmoset. Marmosets are a species of new-world monkeys that split off from our ancestral line before we evolved the ARHGAP11B gene.

If you've ever seen the *Planet of the Apes* movies, you're probably wondering whether inserting a gene that increases mental capacity in monkeys is a good idea. To put aside those worries, the scientists inserted the gene into monkey embryos, but ended the experiments before the animals were born.

The study found that placing the gene in the monkeys did increase the number of pre-neuron cells in their brain, in an area called the outer subventricular zone. It also increased the number of upper-layer neurons. This made the cortex bigger and even caused it to start wrinkling, which doesn't normally happen in marmosets. In other words, the experiment confirmed that having this gene does scale up the size of your brain. Did it make the monkeys smarter? The animals were not allowed to be born, so we don't know. But later studies repeated the experiment with mice, and this time the embryos were allowed to be born. The scientists found that the altered mice had bigger brains and also greater flexibility in their memory system, as well as less anxiety.

Another genetic clue is a gene called SRGAP2. Scientists believe this gene causes the brain to solidify or freeze the connections between neurons as it grows. Most apes have one copy of this gene, but humans have three. What do these extra copies do? The extra copies seem to slow down the activity of the first copy of the gene, giving the brain more time to make connections before they are set in stone. This could be why it takes so long for human brains to develop—more than twenty years to fully mature—while with other species like monkeys it takes only around four years.

This delayed maturation could be part of what gives humans their advantage. By extending the time neurons have to make connections, the brain is able to make deeper and more complicated computational networks.

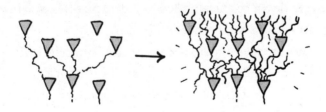

Finally, we have to consider the possibility that what gives humans their advantage isn't necessarily a bigger or better-connected brain, but just having the right upgrades. Scientists think there are two specific brain circuits that humans developed that could explain our accelerated advancement: tools and symbolic language.

Tool use is more common in the animal kingdom than you might think. Animals as diverse as otters, birds, and dolphins have been seen using tools like rocks and sticks. Even some fish are reported to use rocks as crude anvils. Chimpanzees, in particular, are able to use small sticks to extract termites from mounds, to dig in dirt to get to food, and to crack nuts. They are also the only nonhuman animal that use tools as weapons. They've been seen sharpening sticks with their teeth to use as spears for hunting smaller animals.

I call it the Ant-o-Matic 3000.

The common view among scientists is that there is no clean divider between humans and other animals when it comes to using tools. Tool use seems to be a spectrum, with simple tool "users" at one end, and tool "makers and designers" at the other. It's clear, though, that humans sit at the far end of the spectrum when it comes to making and designing tools. After all, our tools have been able to see down to the atomic level, take pictures of faraway black holes, and cure a wide range of diseases.

The advantage humans seem to have is in refining this ability, and this may be due to special regions of the cortex we've evolved that expand our capacity for tool use. Some scientists think one such area is the left anterior supramarginal gyrus, or aSMG. How do they know this? In one experiment, the scientists put both humans and monkeys in fMRI machines and showed them videos of hands grabbing objects, and also videos of hands using tools like a screwdriver or a pair of pliers.

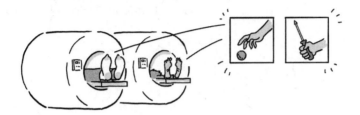

They found that humans and monkeys responded almost the same way when watching both kinds of videos. But when watching the videos of hands using tools, an extra brain region lit up in the human subjects' brains: the aSMG. Monkeys don't seem to have this special area. The scientists think this extra brain region lets humans understand more deeply the connection between using tools and what happens when you use them.

The second important brain upgrade humans seem to have is language processing, and specifically our ability to use symbols to communicate.

Many animal species communicate, either through calls, grunts, or whistles. Humans use these too, but we also communicate with symbols, like these letters you are reading. The ability to use symbolic language sets us apart from all other creatures on earth. Sure, other animals can be taught to understand and use symbols, like the famous experiments with the chimpanzee Washoe, or the lowland gorilla Koko. But as far as we know, no other animals are able to write down what they want to say.

Without this symbolic language ability, we wouldn't have any way to transmit information beyond our own existence. Books, comics, music, and even languages might be lost to us without it. Just think, it's written language that's allowed us to accumulate and refine knowledge. We can share ideas, good and bad, around the world and across time with this skill.

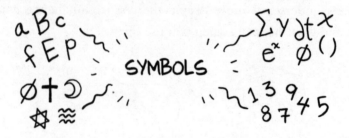

Where in the brain can we find this ability? As with the use of tools, there are some genetic suspects when it comes to advanced language.

The gene that makes a protein called forkhead box protein P2 (FOXP2) was one of the first genes proposed to be important in our language skills. It was originally discovered by studying a human family that had the same problem with speech and language over three generations (i.e., the child had it, the parent had it, and the grandparent had it). An analysis of their genes pointed the finger at FOXP2 as the culprit.

FOXP2

FOXP2 encodes a protein that is used a lot in the brain during the development of the fetus. At first, studies seemed to say that only early *Homo sapiens* had this gene, but then later it was found in Neanderthals and Denisovans as well. The gene was also later found in mice, where it seems to be important for their ability to vocalize sounds. So, FOXP2 is probably not the only culprit in giving us advanced language skills, which makes sense since language is a very complex task.

What Really Makes Us Human

One of the first encounters many of us have with the idea of evolution is the famous 1965 infographic by natural history painter Rudolph Zallinger, called "The March of Progress." You've probably seen this version:

It's an effective infographic that tells a compelling story about change from one species to another. But it's a lie.

The first and most obvious problem is that there are no women in the original graphic, just men. And, if you take it literally, you might think that human evolution was a steady path from apes to modern humans. Indeed, its very name—"The March of Progress"—gives the impression that what's on the right is better or more highly evolved than what's on the left. For many reasons, that's not quite right.

Humans tend to have an overinflated opinion of who we are. There's a concept in psychology called the "end-of-history illusion" that describes how people tend to think of how they are today as the best version of themselves. And not only that, they think that they are the best version of themselves that they'll *ever* be. You tend to think this when you're young, and you still think that when you're old.

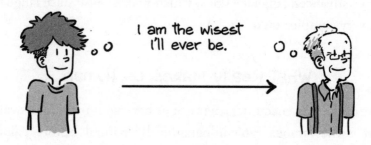

I am the wisest I'll ever be.

Along the same lines, we propose there is also an "end-of-evolution illusion." It's the conceit that humans are the apex of evolution, and that, like "The March of Progress" implies, the human brain is the optimal end product of billions of years of genetic tinkering.

In reality, evolution is more like a sprawling tree, and we are simply perched on one branch of that tree. For example, it's commonly misunderstood that we evolved from apes—but that's like thinking that you evolved from your cousin. What actually happened is that both apes and humans have a common ancestor (now extinct) from which we both evolved millions of years ago.

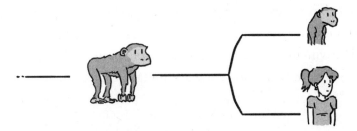

It's also false to assume that humans have "won" the genetic race. If we define "winning" as having the most biomass on the planet, it's not even close; plants are more successful by this measure, followed by bacteria. Your own body plays host to nearly 39 trillion bacteria! There are 5,000 times more bacteria just in your body than there are human beings on the planet.

It's a hard and humbling truth that having a smart brain is not essential to survive and thrive over the lifespan of planet Earth. Horseshoe crabs, which evolved 445 million years ago, are still around today, and they are doing just fine with their simple, tiny brains.

When thinking about what makes us human, it's easy to focus on the differences. And it's natural to emphasize differences that set the rules so that we win the game. But if we put aside this need to win, or to be the best among all species, then it's easier to see what really sets us apart.

Our brains evolved through the process of natural selection, just like all other brains. And through sheer happenstance, we stumbled on a particular skill: the ability to adapt and to solve problems. We skirted by on this skill for thousands of years, and then one day it propelled us forward at an enormously accelerated pace. Suddenly, we found ourselves with fewer basic problems to solve (at least for now), leaving us with extra time and extra brain cycles to do other things, like wonder, and create art, literature, and science.

Perhaps what truly sets us apart from all other species is this: our curiosity, our whimsy, and the desire to understand ourselves, as evidenced by your reading this book right now.

What does the future hold? Well, our ability to invent means that we are also the only species able to transcend our biological limits. We can carry our own habitat with us as we explore the deepest ocean or fly into outer space. We also have the technology to change our own genomes, and those of other creatures. And pushing these boundaries means that we are the only species that can overstep them. Ultimately, what makes us human may be our ability to decide for ourselves the answer to that question.

A BRAINY CONCLUSION

The universe belongs to those who, at least to some degree, have figured it out.

—Carl Sagan

Mind if we say a few final words?

If you've made it this far in the book, then you probably read a good chunk of it, and maybe even liked it. Now think about everything that went into that. Think about how your brain was able to use visual processing (or audio processing if you're listening to this book) to register the words, then use memory and language processing to understand them, and then logic and abstract thinking to make sense of them. If we did our jobs right, we also got you to feel emotions. Maybe we got you to think about why you love, or the things you hate or fear. We might have gotten you to think about your past, present, and future, and even your own mortality.

And all of that happened in your brain. Right here, right now, this experience of you reading or hearing these words, that's all happening in your head. In other words, this book isn't just titled *Out of Your Mind,* it is literally coming out of your mind this instant.

The human mind is an amazing thing.

Carl Sagan's quote above reflects the idea that once you understand how something works, there's no limit to what you can do with it. As we continue to uncover the mysteries of the brain, are we getting closer to understanding ourselves and, in doing so, unlocking our true potential?

The brain is a universe in itself. It contains billions of nerve and supporting cells. On average, neurons make thousands of connections with other neurons, which means there could be hundreds of trillions of ways to rewire your brain, each creating new possibilities for a different you.

THE BRAINI-VERSE

Right away, we face a staggering conundrum: Can the human brain understand *itself*? The very organ we are using to understand

336

ourselves is the very thing we're trying to understand. Is it possible for 86 billion neurons to fully grasp the workings of 86 billion neurons? It seems improbable that we can reach a kind of real truth about our minds.

And yet we have a secret weapon: science. The practice of science is a commitment to measure and refine—to find patterns and come up with models that can explain and predict. From this, we can at least form a picture of how the brain works. In this book, we've tried to tell you the story of how humans have been using science (philosophy, psychology, neuroscience, chemistry, etc.) to answer some of the deepest questions we have about ourselves.

So, what have we learned?

We learned that the human brain had a humble origin. We know that its sparks and shimmers began as simple proteins controlling the amount of water inside cells. And we learned that through billions of years of evolution these cells became something more—an organ capable of contemplating and understanding its own nature.

We've learned that the brain is organized into areas that are specialized, and that also depend on complex interconnections to work properly. We also know that specific brain areas can do more than one thing. For example, we learned that the brain's reward system helps drive our wants and desires. Love, hate, addiction, and other basic motivations all depend on this one system.

Science has also shown us that the brain communicates with itself in a kind of computer code. The brain isn't a machine in a literal sense, but it is a massively parallel electrochemical biological device that does computations. And, unlike a typical computer, its individual components are malleable and can change moment to moment throughout its lifespan.

We also learned how science has tackled the concepts of memory, consciousness, and free will, and found that our sense of being depends on all of these things. To be human is to hold in your head what has happened before, what is happening now, and what we think might happen in the future, all at the same time.

The big picture from all this is that the brain CAN be understood. It is fascinating to think that we can use science to tackle and make sense of the most subjective topic in the world: the experience of being an individual human.

Are we done? Of course not.

As we write this, there is a revolution happening in artificial intelligence (AI). What was a slow rise to smarter and smarter machines seems to have reached its predicted exponential increase. Many of the current AI models are being hailed as leading to "general intelligence." Some scientists are saying that something that resembles "theory of mind" (a sign of consciousness) is already present in some models.

What does that mean for the human mind? If AIs overtake the human brain in computational ability, does it mean that neuroscience is a lost cause, like the study of an obsolete model?

Hopefully not! If anything, the rise of AI makes understanding the human mind even more important. After all, we get to design these AI systems, and the closer we can design them to humans (or to be different from humans in specific ways), the more we can make sure that they are serving us in predictable ways. And to do that, we need brain science.

Clearly, we still have a way to go. Many of the answers in this book are incomplete, evidence that there are still great mysteries at every corner of the human psyche. And you, as one of its users, can be a part of it. The mind remains a great frontier, and we need thinkers and artists to join us in exploring the perplexing cosmos within our heads.

The answers are there, waiting to come out.

ACKNOWLEDGMENTS

We have a good mind to thank a lot of people.

We are grateful to friends and colleagues who reviewed early versions of the manuscript: Suelika Chial, Amy Kim Kibuishi, Paul and Linnaea Scott, Alisha Kamath, Lucas Godwin, Marcia Sullivan, Hamideh Sadat Bagherzadeh, Aqil Izadysadr, and Ed Ergenzinger.

Special thanks to our editor, Denise Oswald, for her excitement for comics and science, her trust in us, and her steady guidance. Thanks to Seth Fishman for always finding the right place for our work. Thanks to the whole team at the Gernert Company, including Rebecca Gardner, Will Roberts, Ellen Goodson Coughtrey, Nora Gonzalez, and Jack Gernert, and to their international counterparts. Many thanks to everyone at Pantheon Books who contributed their time and talent to the making and release of this book.

Jorge is grateful to his family, as always, for their continued support and encouragement.

Dwayne is grateful to his family and friends, for their encouragement and kindness; to his students and postdoctoral fellows who have taught him more than he has taught them; and to mentors and colleagues who lifted him up at key moments in his life.

Most of all, we are grateful to you for reading this book.

Chapter 1: Where Is the Mind?

Breasted JH (1980). *The Edwin Smith Surgical Papyrus.* University of Chicago Press.

Broca P (1861). "Loss of speech, chronic softening and partial destruction of the anterior left lobe of the brain," *Bulletin de la Société Anthropologique* 2, 235–38. (English translation by Christopher D. Green.)

Ferrier D (1876). *The Functions of the Brain.* New York: G. P. Putnam's Sons.

Harlow JM (1848). "Passage of an Iron Rod Through the Head." *Boston Medical and Surgical Journal* 39 (20): 389–93.

Meltzer ES and Sanchez GM (2014). *The Edwin Smith Papyrus: Updated Translation of the Trauma Treatise and Modern Medical Commentaries.* Lockwood Press.

O'Driscoll K and Leach JP (1998). " 'No longer Gage': An iron bar through the head. Early observations of personality change after injury to the prefrontal cortex." *BMJ* 317(7174): 1673–74.

Penfield W and Boldrey E (1937). "Somatic motor and sensory representation in the cerebral cortex of man as studied by electrical stimulation." *Brain* 60: 389–443.

Shelley M (1821). *Frankenstein: or, the Modern Prometheus.* London: Henry Colburn and Richard Bentley. Project Gutenberg.

Wernicke K (1874). *Der aphasische Symptomencomplex. Eine psychologische Studie auf anatomischer Basis* ["The aphasic symptom complex: A psychological study from an anatomical basis"]. Breslau: M. Crohn und Weigert.

Chapter 2: Why Do We Love?

Aron A, Fisher H, Mashek, DJ, Strong, G, Li H, and Brown LL (2005). "Reward, motivation, and emotion systems associated with early-stage intense romantic love." *Journal of Neurophysiology* 94(1): 327–37.

Bartels A and Zeki S (2004). "The neural correlates of maternal and romantic love." *NeuroImage* 21(3): 1155–66.

Burkett JP and Young LJ (2012). "The behavioral, anatomical and pharmacological parallels between social attachment, love and addiction." *Psychopharmacology* 224: 1–26.

Damasio A and Carvalho GB (2013). "The nature of feelings: Evolutionary and neurobiological origins." *Nature Reviews Neuroscience* 14: 143–52.

Hatfield E, Bensman L, and Rapson RL (2012). "A brief history of social scientists' attempts to measure passionate love." *Journal of Social and Personal Relationships* 29 (2): 143–64.

Jankowiak WR and Fischer EF (1992). "A cross-cultural perspective on romantic love." *Ethnology* 31(2): 149–55.

Lee H-J, Macbeth AH, Pagani J, and Young WS (2009). "Oxytocin: The Great facilitator of life." *Progress in Neurobiology* 88(2): 127–51.

Marsh N, Marsh AA, Lee MR, and Hurlemann R (2021). "Oxytocin and the neurobiology of prosocial behavior." *The Neuroscientist* 27(6): 604–19.

Poldrack R (2006). "Can cognitive processes be inferred from neuroimaging data?" *Trends in Cognitive Sciences* 10 (2): 59–63.

Sobota R, Mihara T, Forrest A, Featherstone RE, and Siegel SJ (2015). "Oxytocin reduces amygdala activity, increases social interactions, and reduces anxiety-like behavior irrespective of NMDAR antagonism." *Behavioral Neuroscience* 129(4): 389–98.

Chapter 3: Why Do We Hate?

Glidden J, D'Esterre A, and Killen M (2021). "Morally-relevant theory of mind mediates the relationship between group membership and moral judgments." *Cognitive Development* 57: 100976.

Harrington ER (2004). "The social psychology of hatred." *Journal of Hate Studies* 3(1): 49–82.

Hein G, Silani G, Preuschoff K, Batson CD, and Singer T (2010). "Neural responses to ingroup and outgroup members' suffering predict individual differences in costly helping." *Neuron* 68(1): 149–60.

Kredlow, AM, Fenster RJ, Laurent ES, Ressler KJ, and Phelps EA (2022). "Prefrontal cortex, amygdala, and threat processing: implications for PTSD." *Neuropsychopharmacology* 47(1): 247–59.

Lasko EN, Dagher AC, West SJ, and Chester DS (2022). "Neural mechanisms of intergroup exclusion and retaliatory aggression." *Social Neuroscience* 17(4): 339–51.

McGlothlin H and Killen M (2006). "Intergroup attitudes of European American children attending ethnically homogeneous schools." *Child Development* 77(5): 1375–86.

Molenberghs P, Bosworth R, Nott Z, Louis WR, Smith JR, Amiot CE, Vohs KD, and Decety J (2014). "The influence of group membership and individual differences in psychopathy and perspective taking on neural responses when punishing and rewarding others." *Human Brain Mapping* 35(10): 4989–99.

Radke S, Volman I, Mehta P, Van Son V, Enter D, Sanfey A, Toni I, De Bruijn ERA, and Roelofs K (2015). "Testosterone biases the amygdala toward social threat approach." *Science Advances* 1(5): e1400074.

Segal H (1974). *Introduction to the Work of Melanie Klein.* New York: Basic Books.

Simi P, Blee K, DeMichele M, and Windisch S (2017). "Addicted to hate: Identity residual among former white supremacists." *American Sociological Review* 82(6): 1167–87.

Sternberg RJ (2003). "A duplex theory of hate: Development and application to terrorism, massacres, and genocide." *Review of General Psychology* 7(3): 299–328.

Tiihonen J, Rautiainen MR, Ollila HM, Repo-Tiihonen E, Virkkunen M, Palotie A, Pietiläinen O, Kristiansson K, Joukamaa M, Lauerma H, Saarela J, Tyni S, Vartiainen H, Paananen J, Goldman D, and Paunio T (2015). "Genetic background of extreme violent behavior." *Mol Psychiatry* 20(6): 786–92.

Weinstein N, Ryan WS, DeHaan CR, Przybylski AK, Legate N, and Ryan RM (2012). "Parental autonomy support and discrepancies between implicit and explicit sexual identities: Dynamics of self-acceptance and defense." *Journal of Personality and Social Psychology* 102(4): 815–32.

White SF, Lee Y, Phan JM, Moody SN, and Shirtcliff EA (2019). "Putting the flight in 'fight-or-flight': Testosterone reactivity to skydiving is modulated by autonomic activation." *Biol Psychol.* 143 (April 2019): 93–102.

Zeki S and Romaya JP (2008). "Neural correlates of hate." *PLoS One* 3(10): e3556.

Comic Interlude: A Primer on Fear

Digdon N (2020). "The Little Albert controversy: Intuition, confirmation bias, and logic." *Hist Psychol.* 23: 122–131.

Powell RA and Schmaltz RM (2021). "Did Little Albert actually acquire a conditioned fear of furry animals? What the film evidence tells us." *Hist Psychol.* 24: 164–81.

Tovote P, Fadok J, and Lüthi A (2015). "Neuronal circuits for fear and anxiety." *Nat Rev Neurosci* 16: 317–31.

Chapter 4: Will an AI Take My Job?

Glickstein M (2006). "Golgi and Cajal: The neuron doctrine and the 100th anniversary of the 1906 Nobel Prize." *Current Biology* 16(5): R147–R151.

Golgi, C (1885). Sulla fina anatomia degli organi centrali del sistema nervoso. Reggio-Emilia: *S. Calderini e Figlio;* 1885.

Golgi C (1906). "The neuron doctrine—theory and facts." Nobel Lecture. Nobel Prize.org.

Kang HW, Kim HK, Moon BH, Lee SJ, and Rhyu IJ (2017). "Comprehensive review of Golgi staining methods for nervous tissue." *Applied Microscopy* 47(2): 63–69.

Pannese, E (1999). "The Golgi stain: invention, diffusion and impact on neurosciences." *Journal of the History of the Neurosciences* 8(2): 132–40.

Ramón y Cajal S (1906). "The structure and connexions of neurons." Nobel Lecture. NobelPrize.org.

Sherrington CS (1906). *The integrative action of the nervous system.* Yale University Press.

von Bartheld CS, Bahney J, and Herculano-Houzel S (2016). "The search for true numbers of neurons and glial cells in the human brain: A review of 150 years of cell counting." *J Comp Neurol* 524(18): 3865–95.

Young NA, Collins CE, and Kaas JH (2013). "Cell and neuron densities in the primary motor cortex of primates." *Front Neural Circuits* 7: 30.

Chapter 5: What Are the Limits of Memory?

Abraham WC, Jones OD, and Glanzman DL (2019). "Is plasticity of synapses the mechanism of long-term memory storage?" *NPJ Science of Learning* 4(1): 9.

Annese J, Schenker-Ahmed NM, Bartsch H, Maechler P, Sheh C, Thomas N, Kayano J, Ghatan A, Bresler N, Frosch MP, Klaming R, and Corkin S (2014). "Postmortem examination of patient H.M.'s brain based on histological sectioning and digital 3D reconstruction." *Nat Commun* 5: 3122.

Corkin S (2013). *Permanent Present Tense: The Unforgettable Life of the Amnesic Patient, H.M.* Basic Books.

Cowan N (2012). *Working memory capacity.* Psychology Press.

Hennig MH (2013). "Theoretical models of synaptic short term plasticity." *Front Comput Neurosci* 7: 45.

Hirano T (2013). "Long-term depression and other synaptic plasticity in the cerebellum." *Proc Jpn Acad Ser B Phys Biol Sci* 89(5): 183–95.

Jabr F (2011). "Cache cab: Taxi drivers' brains grow to navigate London's streets." *Scientific American*. https://www.scientificamerican.com/article/london-taxi-memory/. Accessed 5/5/2024.

Ma WJ, Husain M, and Bays PM (2014). "Changing concepts of working memory." *Nature Neuroscience* 17(3): 347–56.

Maguire EA, Gadian DG, Johnsrude IS, Good CD, Ashburner J, Frackowiak RSJ, and Frith CD (2000). "Navigation-Related Structural Change in the Hippocampi of Taxi Drivers." *Proceedings of the National Academy of Sciences* 97(8): 4398–4403.

Markowitsch HJ and Staniloiu A (2023). "Behavioral, neurological, and psychiatric frailty of autobiographical memory." *Wiley Interdisciplinary Reviews: Cognitive Science* 14(3): e1617.

Miller GA (1956). "The magical number seven, plus or minus two: Some limits on our capacity for processing information." *Psychological Review* 63(2): 81–97.

Moser M-B, Rowland DC, and Moser EI (2015). "Place cells, grid cells, and memory." *Cold Spring Harbor Perspectives in Biology* 7(2): a021808.

O'Keefe J and Dostrovsky J (1971). "The hippocampus as a spatial map. Preliminary evidence from unit activity in the freely-moving rat." *Brain Research* 34(1): 171–75.

Parker ES, Cahill L, and McGaugh JL (2006). "A case of unusual autobiographical remembering." *Neurocase* 12(1): 35–49.

Purves D, Augustine GJ, Fitzpatrick D, et al., eds. (2001). "Mechanisms of Short-Term Synaptic Plasticity in the Mammalian Nervous System." *Neuroscience,* 2nd ed. Sunderland (MA): Sinauer Associates.

Riegel DC (2020). "Discovering memory: Using sea slugs to teach learning and memory." *J Undergrad Neurosci Educ* 19(1): R19–R22.

Rosenblum Y and Dresler M (2021). "Can brain stimulation boost memory performance?" *PLOS Biology* 19(9): e3001404.

Scoville WB (1968). "Amnesia after bilateral mesial temporal-lobe excision: Introduction to case H.M." *Neuropsychologia* 6: 211–13.

Scoville WB and Milner B (1957). "Loss of recent memory after bilateral hippocampal lesions." *J Neurol Neurosurg Psychiatry* 20: 11–21.

Treffert DA (2009). "The savant syndrome: an extraordinary condition. A synopsis: past, present, future." *Philosophical Transactions of the Royal Society B: Biological Sciences* 364(1522): 1351–57.

Wilson MA and McNaughton BL. "Reactivation of hippocampal ensemble memories during sleep." *Science* 265: 676–79.

Xia C. (2006). "Understanding the human brain: A lifetime of dedicated pursuit. Interview with Dr. Brenda Milner." *McGill Journal of Medicine* 9(2): 165.

Young RL, Ridding MC, and Morrell TL (2004). "Switching skills on by turning off part of the brain." *Neurocase* 10(3): 215–22.

Zhang J (2019). "Basic neural units of the brain: neurons, synapses and action potential." ArXiv preprint arXiv: 1906.01703.

Comic Interlude: A Primer on Alzheimer's

Alzheimer A, Stelzmann RA, Schnitzlein HN, and Murtagh FR (1995). "An English translation of Alzheimer's 1907 paper, Über eine eigenartige Erkankung der Hirnrinde." *Clin Anat.* 8: 429–31.

Frisoni, GB, Altomare D, Thal DR, et al. (2022). "The probabilistic model of Alzheimer disease: The amyloid hypothesis revised." *Nat Rev Neurosci* 23: 53–66.

Grieco SF, Holmes TC, and Xu X (2023). "Probing neural circuit mechanisms in Alzheimer's disease using novel technologies." *Mol Psychiatry* 28: 4407–20.

Chapter 6: What Is Addiction?

The graph of common addictive substances plotted according to their harmfulness and addictiveness is redrawn from: Nutt D, King LA, Saulsbury W, and Blakemore C (2007). Development of a rational scale to assess the harm of drugs of potential misuse. *Lancet* 369: 1047–53.

Akpan, Nslkan. "Fentanyl Is So Potent Doctors Don't Know How to Fight It," December 2017. https://www.pbs.org/newshour/science/fentanyl-is-so-potent-doctors-dont-know-how-to-fight-it. Accessed 5/5/2024.

Andreas P (2020). *Killer High: A History of War in Six Drugs.* Oxford University Press.

Anselme P and Robinson MJ (2013). "What motivates gambling behavior? Insight into dopamine's role." *Front Behav Neurosci* 7: 182.

Antonio A, Brennan A, and Conversi D (2021). "The SEEKING drive and its fixation: A neuro-psycho-evolutionary approach to the pathology of addiction." *Front Hum Neuroscien* 15: 635932.

Berridge KC and Kringelbach ML (2015). "Pleasure systems in the brain." *Neuron* 86 (3): 646–64.

Cash H, Rae CD, Steel AH, and Winkler A (2012). "Internet addiction: a brief summary of research and practice." *Curr Psychiatry Rev* 8(4): 292–98.

Everitt BJ, Hutcheson DM, Ersche KD, Pelloux Y, Dalley JW, and Robbins TW. "The orbital prefrontal cortex and drug addiction in laboratory animals and humans." *Annals of the New York Academy of Sciences* 1121(1): 576–97.

James A and Williams J (2020). "Basic opioid pharmacology—an update." *British Journal of Pain* 14(2): 115–21.

Lisman JE and Grace AA (2005). "The hippocampal-VTA loop: Controlling the entry of information into long-term memory." *Neuron* 46(5): 703–13.

Maslow AH (1943). "A theory of human motivation." *Psychological Review* 50(4): 370–96.

Pahuja R, Seth K, Shukla A, Shukla RK, Bhatnagar P, Chauhan LK, Saxena PN, Arun J, Chaudhari BP, Patel DK, Singh SP, Shukla R, Khanna VK, Kumar P, Chaturvedi RK, and Gupta KC. (2015). "Trans-blood brain barrier delivery of dopamine-loaded nanoparticles reverses functional deficits in parkinsonian rats." *ACS Nano* 9(5): 4850–71.

Rhodes T, Lilly R, Fernández C, Giorgino E, Kemmesis UE, Ossebaard HC, Lalam N, Faasen I, and Spannow KE (2003). "Risk factors associated with drug use: The importance of 'risk environment.'" *Drugs: Education, Prevention and Policy* 10(4): 303–29.

Roque Bravo R, Faria AC, Brito-da-Costa AM, Carmo H, Mladěnka P, Dias da Silva D, Remião F On behalf of the oemonom researchers (2022). "Cocaine: An updated overview on chemistry, detection, biokinetics, and pharmacotoxicological aspects including abuse pattern." *Toxins (Basel)* 14(4): 278.

Voon V, Mole TB, Banca P, Porter L, Morris L, Mitchell S, Lapa TR, Karr J, Harrison NA, Potenza MN, and Irvine M (2014). "Neural correlates of sexual cue reactivity in individuals with and without compulsive sexual behaviours." *PLoS One* 9(7): e102419.

Weinstein A and Lejoyeux M (2015). "New developments on the neurobiological and pharmaco-genetic mechanisms underlying internet and videogame addiction." *American Journal on Addictions* 24(2): v117–25.

Comic Interlude: The Case of the Addicted Detective

Conan Doyle A DPS (1937). "Was Sherlock Holmes a drug addict?" *The Lancet,* 229: 292.

Editorial (1937). "Was Sherlock Holmes a drug addict?" *Nature* 139: 21.

Lüscher C, Robbins TW, and Everitt BJ (2020). "The transition to compulsion in addiction." *Nat Rev Neurosci* 21: 247–63.

Martin C. (2010). "Elementary, Dr. Bell." *The Lancet,* 375: 190.

Nestler EJ (2005). "The neurobiology of cocaine addiction." *Sci Pract Perspect.* 3: 4–10.

Chapter 7: What Is Consciousness?

Baars BJ, Geld N, and Kozma R (2021). "Global workspace theory (GWT) and prefrontal cortex: Recent developments." *Frontiers in Psychology* 12: 749868.

Cochrane T (2021). "A case of shared consciousness." *Synthese* 199: 1019–37.

de Haan EHF, Corballis PM, Hillyard SA, Marzi CA, Seth A, Lamme VAF, Volz L, Fabri M, Schechter E, Bayne T, Corballis M, and Pinto Y (2020). "Split-brain: What we know now and why this is important for understanding consciousness." *Neuropsychol Rev* 30(2): 224–33.

Dennett DC (2018). "Facing up to the hard question of consciousness." *Philosophical Transactions of the Royal Society B: Biological Sciences* 373: 20170342.

Green CD (2019). "Where did Freud's iceberg metaphor of mind come from?" *History of Psychology* 22(4): 369–72.

Luppi AI, Craig MM, Pappas I, Finoia P, Williams GB, Allanson J, Pickard JD, Owen AM, Naci L, Menon DK, and Stamatakis EA (2019). "Consciousness-specific dynamic interactions of brain integration and functional diversity." *Nat Commun* 10(1): 4616.

Moutoussis K and Zeki S (2002). "The relationship between cortical activation and perception investigated with invisible stimuli." *Proc Natl Acad Sci USA* 99(14): 9527–32.

Pinto Y, De Haan EHF, and Lamme VAF (2017). "The split-brain phenomenon revisited: A single conscious agent with split perception." *Trends in Cognitive Sciences* 21: 835–51.

Ruch S, Züst MA, and Henke K (2016). "Subliminal messages exert long-term effects on decision-making." *Neuroscience of Consciousness.* 2016 e-collection: niw013.

Searle J (2013). "Theory of mind and Darwin's legacy." *Proceedings of the National Academy of Sciences* 110: 10343–48.

van der Bles AM, Postmes T, and Meijer RR (2015). "Understanding collective discontents: A psychological approach to measuring zeitgeist." *PLoS One* 10(6): e0130100.

van Erp WS, Lavrijsen JC, and Koopmans RT (2016). "The unresponsive wakefulness syndrome: Dutch perspectives." *Nederlands Tijdschrift Voor Geneeskunde* 160: D108.

van Gaal S, de Lange FP, and Cohen MX (2012). "The role of consciousness in cognitive control and decision making." *Front Hum Neurosci* 6: 121.

Weinberger J and Westen D (2008). "RATS, we should have used Clinton: Subliminal priming in political campaigns." *Political Psychology* 29(5): 631–51.

Whalen PJ, Rauch SL, Etcoff NL, McInerney SC, Lee MB, and Jenike MA (1998).

"Masked presentations of emotional facial expressions modulate amygdala activity without explicit knowledge." *J Neurosci* 18(1): 411–18.

Chapter 8: What Makes Us Happy?

Botti S, Orfali K, and Iyengar SS (2009). "Tragic choices: Autonomy and emotional responses to medical decisions." *Journal of Consumer Research* 36 (3): 337–52.

Bouchard TJ, Lykken DT, McGue M, Segal NL, and Tellegen A (1990). "Sources of human psychological differences: The Minnesota study of twins reared apart." *Science* 250: 223–28.

Chavez EJ (2008). "Flow in sport: A study of college athletes." *Imagination, Cognition and Personality* 28(1): 69–91.

Dejonckheere E, Rhee JJ, Baguma PK, Barry O, Becker M, Bilewicz M, Castelain T, Costantini G, Dimdins G, Espinosa A, Finchilescu G, Friese M, Gastardo-Conaco MC, Gómez A, González R, Goto N, Halama P, Hurtado-Parrado C, Jiga-Boy GM, Karl JA, Novak L, Ausmees L, Loughnan S, Mastor KA, McLatchie N, Onyishi IE, Rizwan M, Schaller M, Serafimovska E, Suh EM, Swann WB Jr., Tong EMW, Torres A, Turner RN, Vinogradov A, Wang Z, Yeung VW, Amiot CE, Boonyasiriwat W, Peker M, Van Lange PAM, Vauclair CM, Kuppens P, and Bastian B (2022). "Perceiving societal pressure to be happy is linked to poor well-being, especially in happy nations." *Sci Rep* 12(1): 1514.

Fox GR, Kaplan J, Damasio H, and Damasio A (2015). "Neural correlates of gratitude." *Front Psychol* 6: 1491.

Gao L, Sun B, Du Z, and Lv G (2022). "How wealth inequality affects happiness: The perspective of social comparison." *Frontiers in Psychology* 13: 829707.

Hovorka M, Ewing D, and Middlemas DS (2022). "Chronic SSRI treatment, but not norepinephrine reuptake inhibitor treatment, increases neurogenesis in juvenile rats." *International Journal of Molecular Sciences* 23: 6919.

Kahneman D and Deaton A (2010). "High income improves evaluation of life but not emotional well-being." *Proc Natl Acad Sci USA* 107(38): 16489–93.

Killingsworth MA (2021). "Experienced well-being rises with income, even above $75,000 per year." *Proceedings of the National Academy of Sciences* 118(4): e2016976118.

Kim ES, Whillans AV, Lee MT, Chen Y, and VanderWeele TJ (2020). "Volunteering and subsequent health and well-being in older adults: An outcome-wide longitudinal approach." *American Journal of Preventive Medicine* 59(2): 176–86.

Layard R, Mayraz G, and Nickell S (2010). "Does relative income matter? Are the critics right?" In Diener E, Helliwell JF, and Kahneman D (eds.), *International Differences in Well-Being* (pp. 139–65). Oxford University Press.

Lykken D and Tellegen A (1996). "Happiness is a stochastic phenomenon." *Psychological Science* 7(3): 186–89.

Maslow AH (1943). "A theory of human motivation." *Psychological Review* 50(4): 370–96.

Piff PK and Moskowitz JP (2018). "Wealth, poverty, and happiness: Social class is differentially associated with positive emotions." *Emotion* 18(6): 902–5.

Quello SB, Brady KT, and Sonne SC (2005). "Mood disorders and substance use disorder: A complex comorbidity." *Sci Pract Perspect* 3(1): 13–21.

Reutskaja E, Lindner A, Nagel R, Anderson RA, and Camerer CF (2018). "Choice overload reduces neural signatures of choice set value in dorsal striatum and anterior cingulate cortex." *Nat Hum Behav* 2: 925–35.

Waldinger R (2023). "What makes a good life? Lessons from the longest study on happiness." https://www.ted.com/talks/robert_waldinger_what_makes_a _good_life_lessons_from_the_longest_study_on_happiness?subtitle=en. Accessed 5/5/2024.

Wrigley WJ and Emmerson SB (2013). "The experience of the flow state in live music performance." *Psychology of Music* 41(3): 292–305.

Chapter 9: Do We Have Free Will?

Arain M, Haque M, Johal L, Mathur P, Nel W, Rais A, Sandhu R, and Sharma S (2013). "Maturation of the adolescent brain." *Neuropsychiatric disease and treatment* 9: 449–61.

Babu KS and Barth FG (1984). "Neuroanatomy of the central nervous system of the wandering spider, *Cupiennius salei* (Arachnida, Araneida)." *Zoomorphology* 104: 344–59.

Berdoy M, Webster JP, and Macdonald DW (2000). "Fatal attraction in rats infected with *Toxoplasma gondii*." *Proceedings of the Royal Society of London. Series B: Biological Sciences* 267: 1591–94.

Collias EC and Collias NE (1964). "The development of nest-building behavior in a weaverbird." *The Auk* 81(1): 42–52.

Corver A, Wilkerson N, Miller J, and Gordus A (2021). "Distinct movement patterns generate stages of spider web building." *Current Biology* 31(22): 4983–97.

Darby RR, Horn A, Cushman F, and Fox MD (2018). "Lesion network localization of criminal behavior." *Proceedings of the National Academy of Sciences* 115(3): 601–6.

Franklin B, Majault, Le Roy, Sallin, Bailly JS, D'Arcet, de Bory, Guillotin JI, and Lavoisier A. (2000). "Report of the commissioners charged by the King with the examination of animal magnetism. 1784." *Int J Clin Exp Hypn* 50(4): 332–63.

Marzullo TC (2017). "The missing manuscript of Dr. Jose Delgado's radio controlled bulls." *Journal of undergraduate neuroscience education* 15(2): R29–R35.

Peper JS, Brouwer RM, Boomsma DI, Kahn RS, and Hulshoff Pol HE (2007). "Genetic influences on human brain structure: A review of brain imaging studies in twins." *Human brain mapping* 28(6): 464–73.

United States Senate Ninety-Fifth Congress First Session (1977). "Project MKULTRA, the CIA's program of research in behavioral modification." Joint Hearing before the Select Committee on Intelligence and the Subcommittee on Health and Scientific Research of the Committee on Human Resources. https://www.intelligence.senate.gov/sites/default/files/hearings/95mkultra.pdf Downloaded 5/5/2024.

Zhao J, Feng C, Wang W, Su L, and Jiao J (2022). "Human SERPINA3 induces neocortical folding and improves cognitive ability in mice." *Cell Discov* 8(1): 124.

Comic Interlude: What Makes Something Funny?

Dunbar RI, Baron R, Frangou A, Pearce E, van Leeuwen EJ, Stow J, Partridge G, MacDonald I, Barra V, and van Vugt M (2012). "Social laughter is correlated with an elevated pain threshold." *Proc Biol Sci* 279(1731): 1161–67.

Franklin RG and Adams RB (2011). "The reward of a good joke: Neural correlates of viewing dynamic displays of stand-up comedy." *Cognitive, Affective, and Behavioral Neuroscience* 11: 508–15.

Fried I, Wilson CL, MacDonald KA, and Behnke EJ (1998). "Electric current stimulates laughter." *Nature* 391: 650.

Harris CR and Christenfeld N (1999). "Can a machine tickle?" *Psychonomic Bulletin & Review* 6: 504–10.

LoSchiavo FM, Shatz MA, and Poling DA (2008). "Strengthening the scholarship of teaching and learning via experimentation." *Teaching of Psychology* 35(4): 301–4.

Parvizi J, Anderson SW, Martin CO, Damasio H, and Damasio AR (2001). "Pathological laughter and crying: A link to the cerebellum." *Brain* 124(9): 1708–19.

Wild B, Rodden FA, Grodd W, and Ruch W (2003). "Neural correlates of laughter and humour." *Brain* 126(10): 2121–38.

Chapter 10: What Happens When We Die?

Anonymous (1907). "Soul has weight, physician thinks." *New York Times,* March 11, 1907, edition: p5. https://www.nytimes.com/1907/03/11/archives /soul-has-weight-physician-thinks-dr-macdougall-of-haverhill-tells.html. Accessed 5/5/2024.

Baker DJ, Wijshake T, Tchkonia T, LeBrasseur NK, Childs BG, Van De Sluis B, Kirkland JL, and Van Deursen JM (2011). "Clearance of P16Ink4a-positive senescent cells delays ageing-associated disorders." *Nature* 479: 232–36.

Becker E (1973). *The Denial of Death.* Free Press.

Borjigin J, Lee U, Liu T, Pal D, Huff S, Klarr D, Sloboda J, Hernandez J, Wang MM, and Mashour GA (2013). "Surge of neurophysiological coherence and connectivity in the dying brain." *Proc Natl Acad Sci USA* 110(35): 14432–37.

Jaijyan DK, Selariu A, Cruz-Cosme R, Tong M, Yang S, Stefa A, Kekich D, Sadoshima J, Herbig U, Tang Q, Church G, Parrish EL, and Zhu H (2022). "New intranasal and injectable gene therapy for healthy life extension." *Proc Natl Acad Sci USA* 119(20): e2121499119.

Johnston MV (1996). "Cellular alterations associated with perinatal asphyxia." In *Report of the Workshop on Acute Perinatal Asphyxia in Term Infants: August 30–31, 1993, Rockville, Maryland* 16 (96): 27.

Klackl J, Jonas E, and Kronbichler M (2014). "Existential neuroscience: Self-esteem moderates neuronal responses to mortality-related stimuli." *Soc Cogn Affect Neurosci* 9(11): 1754–61.

Lavazza A (2021). "'Consciousnessoids': Clues and insights from human cerebral organoids for the study of consciousness." *Neuroscience of Consciousness* 7(2): niab029.

Lee J, Bignone PA, Coles LS, Liu Y, Snyder E, and Larocca D (2020). "Induced pluripotency and spontaneous reversal of cellular aging in supercentenarian donor cells." *Biochem Biophys Res Commun* 525(3): 563–69.

Parent B and Turi A (2020). "Death's troubled relationship with the law." *AMA Journal of Ethics* 22(12): 1055–61.

Quirin M, Loktyushin A, Arndt J, Küstermann E, Lo YY, Kuhl J, and Eggert L (2012). "Existential neuroscience: a functional magnetic resonance imaging investigation of neural responses to reminders of one's mortality." *Soc Cogn Affect Neurosci* 7(2): 193–98.

Takahashi K and Yamanaka S (2012). "Induction of pluripotent stem cells from mouse embryonic and adult fibroblast cultures by defined factors." *Cell* 126(4): 663–76.

Yenari M and Han H (2012). "Neuroprotective mechanisms of hypothermia in brain ischaemia." *Nat Rev Neurosci* 13: 267–78.

Chapter 11: What Makes Us Human?

The graph that plots the brain size of different animals against the size of their bodies is redrawn from: Tartarelli G and Bisconti M (2006). "Trajectories and Constraints in Brain Evolution in Primates and Cetaceans." *Human Evolution* 21: 275–87. https://doi-org.wake.idm.oclc.org/10.1007/s11598-006-9027-4.

The graph showing the cranial capacity of our recent evolutionary ancestors is redrawn from: DeSilva JM, Traniello JFA, Claxton AG, and Fannin LD (2021). "When and why did human brains decrease in size? A new change-point analysis and insights from brain evolution in ants." *Front Ecol Evol* 9: 712. 10.3389/fevo.2021.742639.

Dicke U and Roth G (2016). "Neuronal factors determining high intelligence." *Philos Trans R Soc Lond B Biol Sci* 371: 20150180.

Eriksen N and Pakkenberg B (2007). "Total neocortical cell number in the mysticete brain." *The Anatomical Record: Advances in Integrative Anatomy and Evolutionary Biology* 290(1): 83–95.

Furutani R (2008). "Laminar and cytoarchitectonic features of the cerebral cortex in the Risso's dolphin *(Grampus griseus),* striped dolphin *(Stenella coeruleoalba),* and bottlenose dolphin *(Tursiops truncatus)*." *Journal of Anatomy* 213(3): 241–48.

Godwin DW and Masicampo M (2015). "Mind like a sponge: Evolutionary paths to the brain." *Biochem* (London) 37(5): 12–15.

Göhde R, Naumann B, Laundon D, Imig C, McDonald K, Cooper BH, Varoqueaux F, Fasshauer D, and Burkhardt P (2021). "Choanoflagellates and the ancestry of neurosecretory vesicles." *Phil Trans R Soc B* 376: 20190759.

Heide M, Haffner C, Murayama A, Kurotaki Y, Shinohara H, Okano H, Sasaki E, and Huttner WB (2020). "Human-specific ARHGAP11B increases size and folding of primate neocortex in the fetal marmoset." *Science* 369: 546–50.

Herculano-Houzel S (2012). "The remarkable, yet not extraordinary, human brain as a scaled-up primate brain and its associated cost." *Proceedings of the National Academy of Sciences* 109(S1): 10661–68.

Hofman MA (2014). "Evolution of the human brain: When bigger is better." *Front Neuroanat* 8: 15.

Hunter P (2020). "The rise of the mammals: Fossil discoveries combined with dating advances give insight into the great mammal expansion." *EMBO Reports* 21(11): e51617.

Kim J, Jung Y, Barcus R, Bachevalier JH, Sanchez MM, Nader MA, and Whitlow CT (2020). "Rhesus macaque brain developmental trajectory: A longitudinal analysis using tensor-based structural morphometry and diffusion tensor imaging." *Cerebral Cortex* 30(8): 4325–35.

Ksepka DT, Balanoff AM, Smith NA, Bever GS, Bhullar BS, Bourdon E, Braun EL,

Burleigh JG, Clarke JA, Colbert MW, Corfield JR, Degrange FJ, De Pietri VL, Early CM, Field DJ, Gignac PM, Gold MEL, Kimball RT, Kawabe S, Lefebvre L, Marugán-Lobón J, Mongle CS, Morhardt A, Norell MA, Ridgely RC, Rothman RS, Scofield RP, Tambussi CP, Torres CR, van Tuinen M, Walsh SA, Watanabe A, Witmer LM, Wright AK, Zanno LE, Jarvis ED, and Smaers JB (2020). "Tempo and pattern of avian brain size evolution." *Curr Biol* 30(11): 2026–36.e3.

Lai CS, Fisher SE, Hurst JA, Vargha-Khadem F, and Monaco AP (2007). "A forkhead-domain gene is mutated in a severe speech and language disorder." *Nature* 413: 519–23.

Marino L, Connor RC, Fordyce RE, Herman LM, Hof PR, Lefebvre L, Lusseau D, McCowan B, Nimchinsky EA, Pack AA, Rendell L, Reidenberg JS, Reiss D, Uhen MD, Van der Gucht E, and Whitehead H (2007). "Cetaceans have complex brains for complex cognition." *PLoS Biol* 5(5): e139.

Olkowicz S, Kocourek M, Lučan RK, Porteš M, Fitch WT, Herculano-Houzel S, and Němec P (2016). "Birds have primate-like numbers of neurons in the forebrain." *Proceedings of the National Academy of Sciences* 113(26): 7255–60.

Orban GA and Caruana F (2014). "The neural basis of human tool use." *Frontiers in Psychology* 5: 81841.

Peeters R, Simone L, Nelissen K, Fabbri-Destro M, Vanduffel W, Rizzolatti G, and Orban GA (2009). "The representation of tool use in humans and monkeys: Common and uniquely human features." *Journal of Neuroscience* 29(37): 11523–39.

Roth G and Dicke U (2012). "Evolution of the brain and intelligence in primates." *Progress in Brain Research* 195: 413–30.

Sherwood CC, Subiaul F, and Zawidzki TW (2008). "A natural history of the human mind: Tracing evolutionary changes in brain and cognition." *J Anat* 212(4): 426–54.

Smaers JB, Rothman RS, Hudson DR, Balanoff AM, Beatty B, Dechmann DKN, de Vries D, Dunn JC, Fleagle JG, Gilbert CC, Goswami A, Iwaniuk AN, Jungers WL, Kerney M, Ksepka DT, Manger PR, Mongle CS, Rohlf FJ, Smith NA, Soligo C, Weisbecker V, and Safi K (2021). "The evolution of mammalian brain size." *Sci Adv* 7(18): eabe2101.

Sun T and Hevner RF (2014). "Growth and folding of the mammalian cerebral cortex: From molecules to malformations." *Nature Reviews Neuroscience* 15(4): 217–32.

A Brainy Conclusion

Sagan C (1979). *Broca's Brain: Reflections on the Romance of Science,* Random House.

Jorge Cham is the Emmy-nominated and bestselling cartoonist creator of the popular online comic strip *Piled Higher and Deeper* (known as PHD Comics—phdcomics.com). He is the cocreator, executive producer, and creative director of the animated PBS Kids show *Elinor Wonders Why* and the author or coauthor of several books about science, including the bestselling and award-winning *We Have No Idea: A Guide to the Unknown Universe* and *Frequently Asked Questions About the Universe,* as well as the children's book series Oliver's Great Big Universe. Cham is also the cohost and cocreator of *Daniel and Jorge Explain the Universe,* an ongoing iHeart Radio podcast. He obtained his PhD in robotics from Stanford University and was an Instructor and Research Associate at Caltech from 2003 to 2005. He is originally from Panama.

Dwayne Godwin is a neuroscientist, an educator, and an academic leader who is a professor in the Department of Translational Neuroscience and served as graduate dean at the Wake Forest University School of Medicine in Winston-Salem, North Carolina. His award-winning research uses techniques ranging from cellular neurophysiology to advanced human neuroimaging in order to better understand the basis of abnormal brain activity, with special interests in calcium channel function, addiction, epilepsy, and traumatic brain injury. The goal of his research is to translate new knowledge into treatments and cures to alleviate the burden of human neurological diseases. His science communication has included *Scientific American Mind* and blogging for the Society for Neuroscience and the Museum of the Moving Image. He received his PhD from the University of Alabama at Birmingham.